Oxford **Mathematics**

Primary Years Programme

5

Contents

NUMBER, PATTERN AND FUNCTION

MEASUREMENT, SHAPE AND SPACE

DATA HANDLING

UNIT 1: TOPIC 1
Place value

Practice

1 What is the value of the red digit in each number?

a 47 625 ________ b 162 752 ________

c 314 520 ________ d 513 804 ________

e 582 400 ________ f 992 008 ________

2 Write the numbers from question 1 in words.

a ________________________________

b ________________________________

c ________________________________

d ________________________________

e ________________________________

f ________________________________

3 Expand the following numbers. The first one has been started for you.

a 32 487:

30 000 + ________________________________

b 316 321:

____

c 40 370:

____

d 806 302:

____

e 450 020:

____

Challenge

1 Look at the digit cards: 2 8 7 9 0 4

a What is the **largest** whole number that can be made using all the cards if the zero is in the thousands place?

____

b What is the **largest** number that can be made if the 9 is in the tens-of-thousands place?

____

c What is the **smallest** number that can be made if the zero is left out?

____

d What is the **largest** number that can be made if the 7 is worth seven-tenths?

____

---

2 Match each number below to the correct spike abacus.

A

B

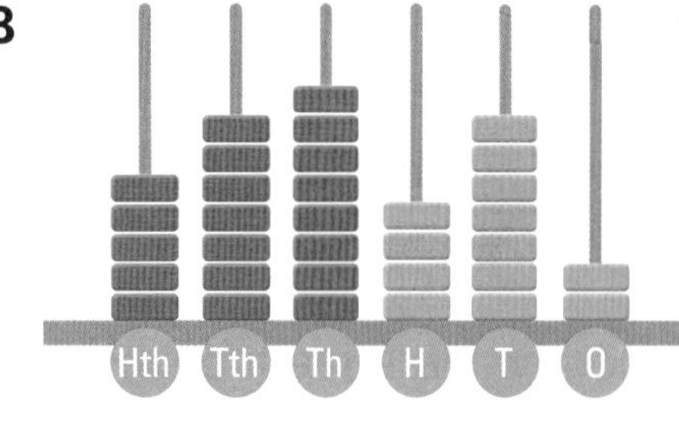

C

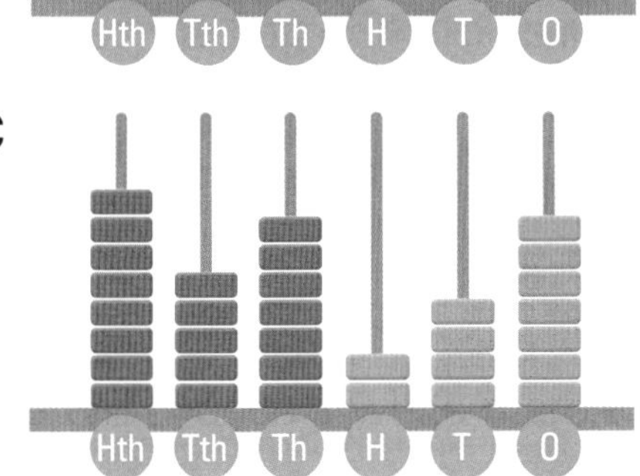

D

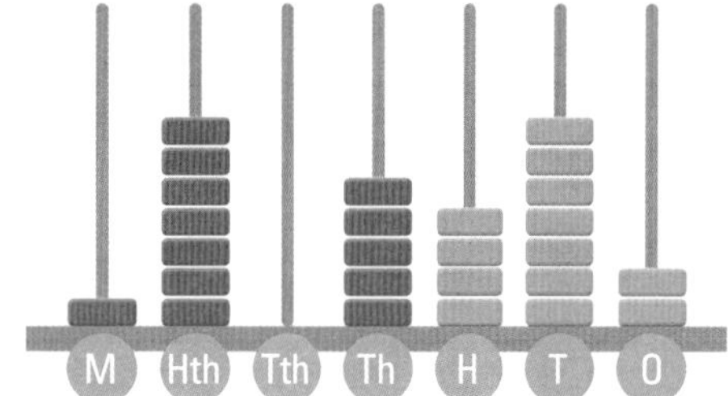

a 578 472 matches to spike abacus __________.

b One million, seven hundred and five thousand, four hundred and seventy-two matches to spike abacus __________.

c 857 427 matches to spike abacus __________.

d Eight hundred and fifty-seven thousand, two hundred and forty-seven matches to spike abacus __________.

---

3 Follow the instructions to round the following numbers. Use digits to write the answers.

a Round 427 899 to the nearest hundred thousand ______________________

b Round 88 699 to the nearest thousand ______________________

c Round 3 743 678 to the nearest ten thousand ______________________

d Round 956 723 to the nearest million ______________________

Mastery

1 This table shows some of the most successful puppet movies ever made. Round the amount of money each movie earned according to the instructions in the table.

Year	Movie	Amount earned in $	Round to the nearest:	Rounded number in $
2011	*The Muppets*	165 184 237	ten thousand	
2014	*Muppets Most Wanted*	80 383 290	million	
1979	*The Muppet Movie*	65 241 000	hundred thousand	
1982	*The Dark Crystal*	40 577 001	hundred thousand	
1996	*Muppet Treasure Island*	34 327 391	ten thousand	
1981	*The Great Muppet Caper*	31 206 251	ten thousand	
1992	*The Muppet Christmas Carol*	27 281 507	thousand	
1984	*The Muppets Take Manhattan*	25 534 703	million	
1999	*Muppets from Space*	22 323 612	hundred thousand	

2 There is a bridge in Taiwan that is 157 317 metres long. This could be rewritten like this: 150 000 + 7000 + 317

Rewrite the length of the bridge in as many ways as you can.

Addition mental strategies

Practice

1 Use the split strategy to answer the following problems. The first one has been started for you.

	Problem	Join the partners	Answer
a	143 + 136	100 + 100 + 40 + 30 + 3 + 6	
b	127 + 331		
c	123 + 456		
d	1324 + 1265		
e	4263 + 3526		

2 Use the compensation (rounding) strategy to answer the following:

a 47 + 98 = ________

b 836 + 201 = ________

c 1243 + 199 = ________

d 8506 + 999 = ________

e 7411 + 2002 = ________

f 4256 + 5999 = ________

3 Use the jump strategy to answer the following:

a 136 + 46 =

b 253 + 86 =

c 1372 + 148 =

d 2578 + 1327 =

4 Use the near-doubles strategy to answer the following:

a 350 + 360 = ________

b 1250 + 1300 = ________

c 5500 + 5600 = ________

d 4950 + 4900 = ________

Challenge

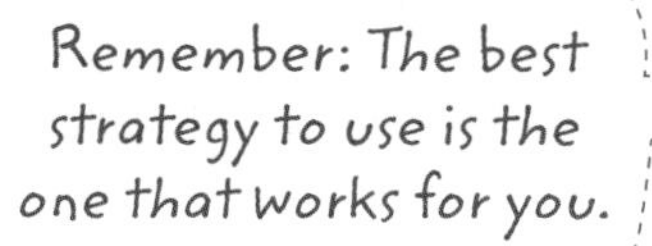

1 Use your choice of strategy to find the answer. Be ready to explain the strategy you used.

a 158 + 39 = ______ **b** 195 + 195 = ______

c 149 + 149 + 149 = ______ **d** 540 + 460 = ______

e 1258 + 1431 = ______ **f** 2950 + 2950 = ______

g 2316 + 2247 = ______ **h** 3502 + 3497 = ______

i 4910 + 5090 = ______ **j** 2718 + 4236 = ______

2 This table lists some of the world's biggest carnivores (creatures that eat other creatures). Decide how to round the numbers. Explain how you have rounded them.

Carnivore	Mass (kg)	Rounded mass (kg)	I rounded to the nearest ...
Blue whale	21 364		
Orca (also known as a killer whale)	9979		
Southern elephant seal	4989		
Walrus	1883		
Steller sea lion	1102		
North American brown bear	783		

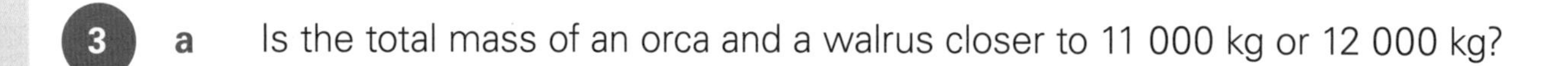

3 **a** Is the total mass of an orca and a walrus closer to 11 000 kg or 12 000 kg?

______

b How much less than 10 000 kg does an orca weigh?

______

c What is the total mass of a walrus and a Steller sea lion?

______

d How much more than 30 000 kg is the total mass of a blue whale and an orca?

______

e The combined mass of which two animals is within 2 kg of the mass of one of the other creatures in the list?

______

f By how much would an orca outweigh two southern elephant seals?

______

Mastery

1 This table lists the most popular sports leagues in the world.

Name and place of league	Type of sport	Average attendance
NFL, USA	American football	68 278
Bundesliga, Germany	Football	43 331
College football, USA	American football	43 288
EPL, England	Football	36 464
AFL, Australia	Australian football	33 428
IPL, India	Cricket	31 750
Major League Baseball, USA	Baseball	30 517
La Liga, Spain	Football	28 498
Big Bash League, Australia	Cricket	28 279

Use mental strategies to solve the following problems. You may find it easier to start by rounding the numbers.

a Is the best estimate for the total of the two cricket crowds 59 000, 60 000 or 61 000? ______________________

b What is the actual total of the two cricket crowds? ______________________

c The total average attendance of the two American football crowds is: __________

2 This table shows the number of vehicles stolen in Australia from September 2016 to September 2017.

State or territory	Number of vehicles stolen	Rounded number
ACT	1193	
NSW	12 209	
NT	1078	
QLD	10 879	
SA	3059	
TAS	1321	
VIC	16 599	
WA	7956	

Source: National Motor Vehicle Theft Reduction Council

a Complete the table by rounding the numbers.

b Make up some back-to-front tasks about the table for others to complete. For example, you could write, "The answer is 2271 vehicles. What is the question?" (The question is, "What is the total number of vehicles stolen in the ACT and the NT?".)

Addition written strategies

Practice

1 Look for patterns in the answers for the following addition problems.

a
```
    1 3 3 0
+     5 8 9
```

b
```
    1 5 4 7
+   1 2 8 1
```

c
```
    2 1 9 1
+   1 5 4 6
```

d
```
    3 0 8 8
+   1 5 5 8
```

e
```
    2 2 0 5
+   5 1 6 8
```

f
```
    3 6 2 3
+   4 6 5 9
```

g
```
    1 3 3 4 8
+     5 8 4 3
```

h
```
    1 2 4 9 9
+   1 5 7 8 3
```

i
```
    1 9 0 5 4
+   1 8 3 1 9
```

j
```
    2 0 6 2 8
+   2 5 8 3 6
```

k
```
    1 8 7 6 9
+   5 4 9 6 8
```

l
```
    2 5 6 5 9
+   5 7 1 6 9
```

m
```
    2 6 6 8 4 6
    5 2 8 4 7 6
+   1 2 3 8 6 9
```

n
```
    1 0 8 5 5 5
    1 5 8 3 6 9
+   5 6 1 3 5 8
```

o
```
    4 5 2 9 6 3
    2 5 8 1 6 5
+     2 6 2 4 5
```

p
```
    1 6 2 4 7 9
    4 5 8 1 9 6
+     2 5 7 8 9
```

q
```
    2 3 8 6 6 4
      8 9 2 4 7
+     4 5 8 2 6
```

r
```
    1 5 7 3 2 9
      6 6 8 8 0
+     5 8 6 1 9
```


OXFORD UNIVERSITY PRESS

Challenge

Remember: Keep the digits in the correct columns.

1 Rewrite the following problems vertically, then solve them.

a 37 183 + 5164 + 863 **b** 38 583 + 14 859 + 879 **c** 823 + 62 090 + 2519

	Tth	Th	H	T	O
+					

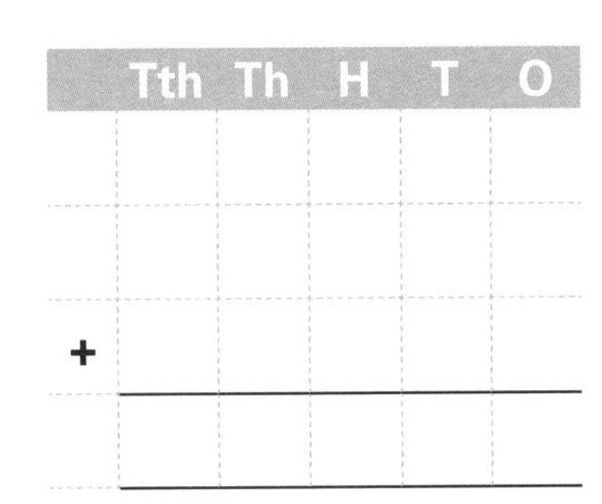

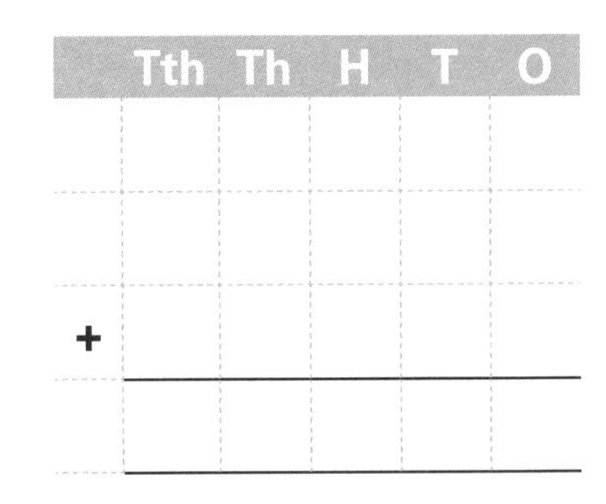

d 98 145 + 995 + 456 923 + 98 258

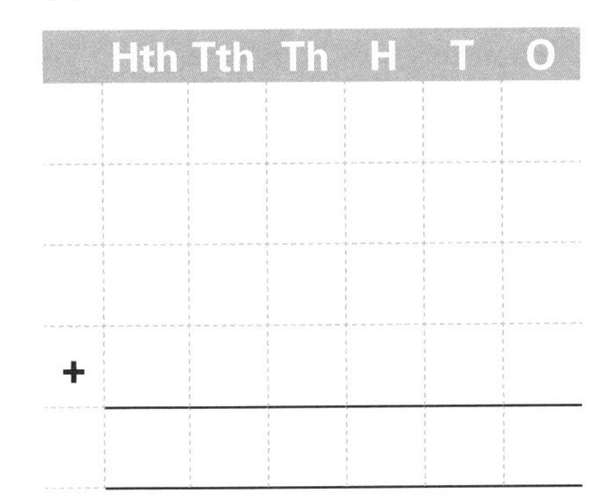

e 525 836 + 23 519 + 5286 + 210 791

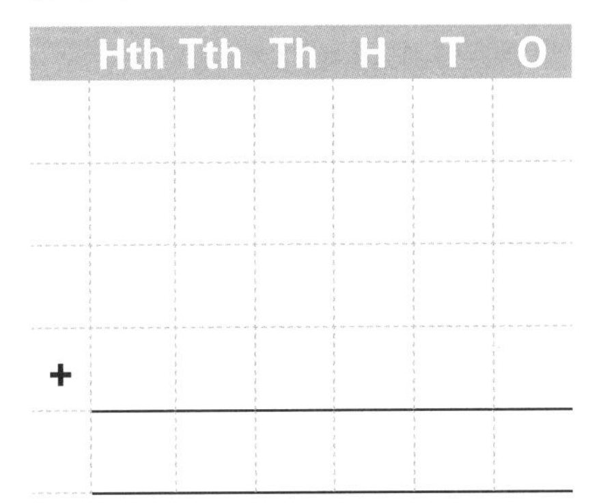

2 The table below shows the area that is covered by trees (and the area that is not covered by trees) in five large countries. The countries are listed in order from the country with the greatest area of trees to the country with the smallest area of trees.

Country	Area with trees	Area without trees	Total area of the country
Brazil	4 935 380 km^2	3 579 497 km^2	
Canada	3 470 690 km^2	6 513 980 km^2	
USA	3 100 950 km^2	6 528 141 km^2	
China	2 083 210 km^2	7 623 751 km^2	
Australia	1 247 511 km^2	6 444 513 km^2	

a Without using a calculator, find the total area of each country. Complete the table.

b The order changes when the total area of each country has been calculated. Reorder the list, from the country with the greatest total area to the one with the smallest.

Mastery

1 a Show the calculator keys pressed to arrive at the answer of 1 000 000.

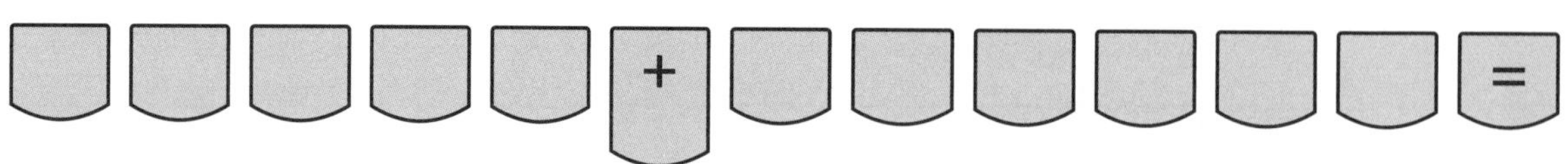

b Write a different addition that arrives at the same answer.

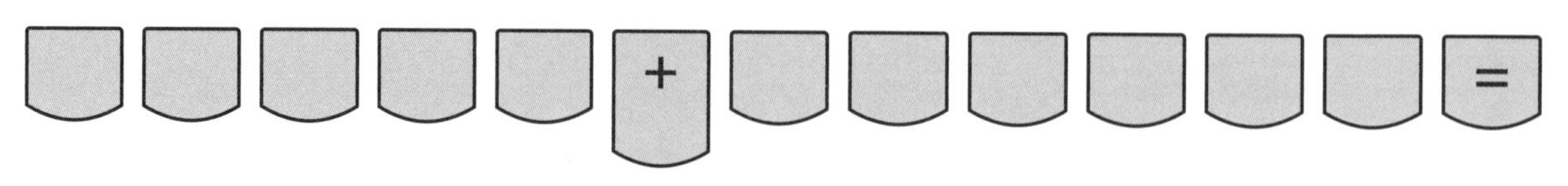

2 If you find the total for the addition below, you will see a pattern in the answer. Make up some three-line additions that continue the pattern. Make sure that at least one line of each addition has a different number of digits to the others.

30 621 + 84 265 + 8570

What are you made of?

3 If someone asked you what you were made of, you would probably say something like, "skin, bones and hair". However, according to scientists it is a lot more boring than that! The average adult is made of the following mix of elements:

- oxygen: 43 000 grams
- carbon: 16 000 grams
- hydrogen: 7000 grams
- nitrogen: 1800 grams
- calcium: 1200 grams
- phosphorus: 780 grams
- sulphur: 140 grams
- potassium: 125 grams
- sodium: 100 grams
- chlorine: 95 grams

If you were to add up all of these elements, would the total equal the mass of an average adult?

Find out what the average adult weighs, and compare this with the total mass of the elements. You could even compare it with your own mass! (You will probably need to convert from grams to kilograms. Use some spare paper for this task.)

Now, try doing some research to find out what other parts of the human body weigh. For example, did you know an adult has more than 10 000 grams of skin?

UNIT 1: TOPIC 4
Subtraction mental strategies

Practice

1 Use the compensation strategy to solve the following problems.

a 76 – 21 = ______

b 483 – 199 = ______

c 427 – 301 = ______

d 1245 – 898 = ______

e 4267 – 2002 = ______

f 7264 – 3998 = ______

2 Use the split strategy to answer the following problems.

a 68 – 35 = ______

b 246 – 132 = ______

c 765 – 534 = ______

d 4227 – 125 = ______

e 7539 – 3435 = ______

f 8694 – 4372 = ______

3 Use the open number line or a mental strategy of your choice to answer the following problems.

a 875 – 450 =

b 1854 – 126 =

c 1372 – 1155 =

d 2548 – 1337 =

Challenge

1 Use the counting-up strategy to work out how much change you would receive if you used a $50 note to buy each of the following items. Be ready to explain how you arrived at your answer.

a a game at $27.50 ____________

b a jumper at $33.75 ____________

c a book at $24.25 ____________

d a toy at $13.35 ____________

e a hat at $22.15 ____________

f a pen at $7.85 ____________

2 Use a mental strategy of your choice to work out the difference between:

a 78 and 34 ____________

b 67 and 388 ____________

c 3245 and 3199 ____________

d 6428 and 4325 ____________

e 2995 and 5389 ____________

f 8750 and 10 000 ____________

3 This is a list of the five longest rivers in the world. Use mental strategies to answer the questions.

River	Place	Length in km	Length in km, rounded to nearest hundred
Amazon	South America	6992	
Nile	Africa	6853	
Yangtze	China	6380	
Mississippi	USA	6275	
Yenisei	Russia	5539	

a Round each length to the nearest hundred. Record your answers in the table.

b How much less than 7000 km is the Nile?

____

c What is the difference between the lengths of the Mississippi and the Amazon?

____

d If you add the rounded lengths of the Yangtze and the Yenisei rivers, is the total length 100 km or 200 km shorter than 12 000 km?

____

Mastery

The tables below list some of the countries with the smallest populations in the world.

Country	Population
Vatican City	842
Niue	1612
Nauru	10 084
Tuvalu	10 640
Palau	17 948

Country	Population
Cook Islands	18 100
San Marino	33 020
Liechtenstein	37 623
Monaco	38 405
South Ossetia	53 532

1 The following are back-to-front questions. The number shown is the difference between two populations. Your task is to write the question.

a The answer is 9242.

____

b The answer is 42 892.

____

c The answer is 152.

____

d The answer is 37 563.

____

e The answer is 19 675.

____

2 Perhaps there is a country in the list whose population is close to that of your town or suburb. You will need to do some research for some of the following activities. Make sure you have your teacher's permission if you search online.

a Find out the population of the town or suburb where you live. ______________

b What is the difference between your town or suburb's population and that of the country in the list with the closest population? ______________

c What is the difference between the number of students in your school and the population of the smallest country in the list? ______________

d Which country on the list has the largest population? Find the difference between that country's population and your country's population. ______________

e By how much is the population of your country bigger than the population of Monaco? ______________

UNIT 1: TOPIC 5
Subtraction written strategies

Practice

1 Subtraction with trading. Look for patterns in the answers.

a

	Th	H	T	O
	2	1	5	5
–		9	3	4

b

	Th	H	T	O
	4	2	1	1
–	1	8	7	9

c

	Th	H	T	O
	6	2	4	1
–	2	7	9	8

d

	Th	H	T	O
	7	1	3	0
–	2	5	7	6

e

	Th	H	T	O
	9	1	5	3
–	3	4	8	8

f

	Th	H	T	O
	9	4	2	5
–	2	6	4	9

g

	Th	H	T	O
	9	5	8	1
–	1	6	9	4

h

	Th	H	T	O
	9	6	8	3
–		6	8	5

i

	Tth	Th	H	T	O
	1	6	1	9	0
–		3	8	6	9

j

	Tth	Th	H	T	O
	3	5	2	8	9
–	1	1	8	5	7

k

	Tth	Th	H	T	O
	5	6	4	3	0
–	2	1	8	8	7

l

	Tth	Th	H	T	O
	6	7	3	4	0
–	2	1	6	8	6

m

	Tth	Th	H	T	O
	9	2	7	4	3
–	3	5	9	7	8

n

	Tth	Th	H	T	O
	9	6	3	7	3
–	2	8	4	9	7

o

	Hth	Tth	Th	H	T	O
	1	6	6	1	7	7
–		4	2	8	5	6

p

	Hth	Tth	Th	H	T	O
	3	9	2	3	2	0
–	1	5	7	8	8	8

q

	Hth	Tth	Th	H	T	O
	8	1	3	4	8	7
–	4	6	7	9	4	4

r

	Hth	Tth	Th	H	T	O
	7	1	6	2	9	2
–	2	5	9	6	3	8

Challenge

1 The following subtraction problems involve trading across more than one column.

a
```
  5 8 2 0 1
- 2 6 0 7 8
-----------

-----------
```

b
```
  8 4 0 0 2
- 4 9 7 6 8
-----------

-----------
```

c
```
  6 0 0 2 3
-   5 6 7 8
-----------

-----------
```

d
```
  9 0 6 0 3
- 2 5 1 4 7
-----------

-----------
```

e
```
  9 0 1 0 1
- 1 3 5 3 4
-----------

-----------
```

f
```
  9 8 0 0 0
- 1 0 3 2 2
-----------

-----------
```

---

2 This table shows the populations of Australia's states and territories in 2017. Use the information to complete the tasks. Write your subtractions in the space below.

Population of Australia in 2017	
New South Wales	7 861 000
Victoria	6 323 600
Queensland	4 928 500
South Australia	1 723 500
Western Australia	2 580 400
Tasmania	520 900
Northern Territory	246 100
Australian Capital Territory	410 300
Australia	24 594 300

Find the difference between:

a the populations of Northern Territory and Australian Capital Territory

____

b the populations of New South Wales and Victoria

____

c the populations of Western Australia and Tasmania

____

d the populations of Queensland and Victoria

____

e the populations of Australia and New South Wales

____

f the combined populations of New South Wales and Victoria, and the combined populations of all the other states and territories.

____

Mastery

1 On the previous page we looked at the population of Australia. The table below shows the countries with the five largest populations in the world. China has the largest population, with more than **1000 million** people.

Another way to say 1000 million is one billion. One billion written as a number is 1 000 000 000.

	Country	Approximate population
1	China	1 330 141 300
2	India	1 173 108 000
3	USA	310 232 800
4	Indonesia	242 968 300
5	Brazil	201 103 300

a What is the difference between the populations of Australia and China?

____

b Find the difference between the populations of some of the top five countries.

____

____

2 In question 1 on page 15, you might have noticed that all the answers were **palindromes**. A palindrome is a word or number that reads the same if you reverse it. Words like "madam" and numbers like 212 are palindromes.

1001 level 3223 refer radar kayak civic 8118 121 noon 7117

Try making some palindrome subtraction problems for your classmates! Here are some steps to follow:

- Start by making up some six-digit palindromes, such as 123 321. These palindrome numbers will become the answers to your subtraction problems.
- Make sure that each line of the subtraction has at least five digits.
- Give the subtraction problem to someone else to see if they can work out your palindrome answer.

UNIT 1: TOPIC 6

Multiplication mental strategies

Practice

1 Multiply each of these by 10.

a 24 ________ b 543 ________ c 7123 ________

d 1.25 m ________ e 4.5 cm ________ f 2.4 m ________

2 Multiply each of these by 100.

a 67 ________ b 23 ________ c 549 ________

d 4.55 m ________ e 2.5 cm ________ f $1.35 ________

3 Multiplying by multiples of 10. Rewrite the problem and find the product.

		× 30	× 40
a	4	$4 \times 30 = 4 \times 3$ tens = ________ tens $4 \times 30 =$	
b	7		
c	9		

4 Here is a mental strategy for multiplying by 15.

		First multiply by 10	Halve it to find × 5	Add the two answers	Multiplication fact
a	24				
b	32				
c	40				
d	50				
e	35				

Challenge

Using double-doubles and double-double-doubles

Problem: What is 35×8?

- 8 is $2 \times 2 \times 2$ (double-double 2)
- $35 \times 2 = 70$. Double $70 = 140$. Double $140 = 280$
- So, $35 \times 8 = 280$

1 Choose the double-double or double-double-double strategy to answer the following:

a What is 25×8?

b What is 26×8?

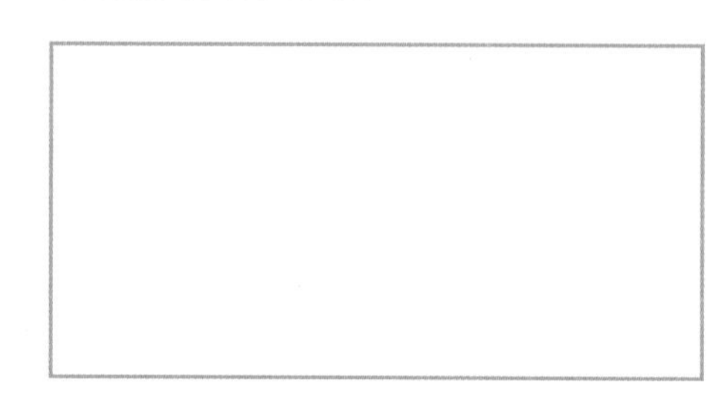

c What is 45×8?

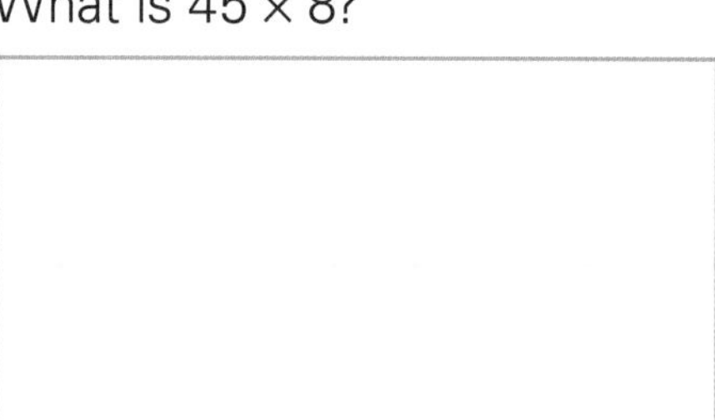

d What is 32×16?

---

2 **a** Jaeger needed 16 lengths of wood to build his cubby house. If each length was 12.5 m long, how many metres did he use altogether? ____________

b Jaeger also bought 36 packets of nails for his cubby house. If there were 20 nails in each packet, how many nails did he buy altogether? ____________

---

3 Write your own word problems to go with each number problem below, then solve them.

a 46×15

____

____

b $\$0.55 \times 40$

____

____

Mastery

1 There are quite a few different mental strategies for multiplication. There is no single strategy that works every time, so it's good to have a range of them to use. Imagine that you are preparing a chart to help your classmates choose a mental strategy. Fill in the gaps in the table below to make the chart as complete as you can.

Multiplication—tips and tricks			
×	**Strategy**	**Example**	**A more difficult example**
× 2	Double the number.	16 × 2 = ? 16 + 16 = 32 So, 16 × 2 = 32	42 × 2 = ? 42 + 42 = 84 So, 42 × 2 = 84
× 3	Double the number, then add the answer to the first number.	15 × 3 = ? 15 × 2 = ______ Answer + 15 = ______	
× 4			
× 5	Multiply by 10, then …		
× 6	Multiply by 3, then …		
× 7	Nobody has thought of a strategy. Can you?		
× 8			
× 9			
× 10			
× 20, × 30, × 40 etc.			

UNIT 1: TOPIC 7
Multiplication written strategies

Practice

1 
Shade the model and fill in the blanks.

a 6 × 46

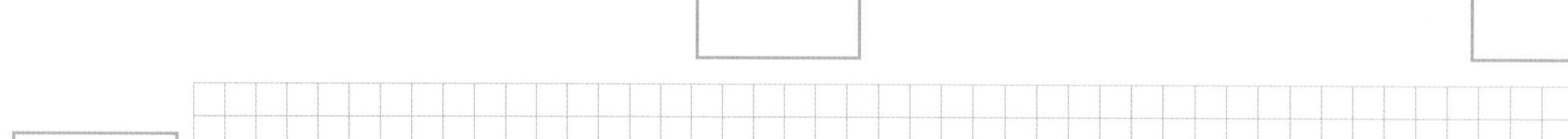

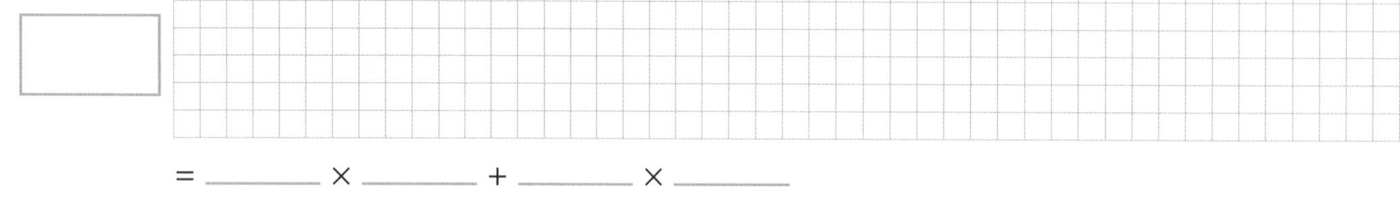

= ______ × ______ + ______ × ______

= ______ + ______

= ______

b 7 × 37

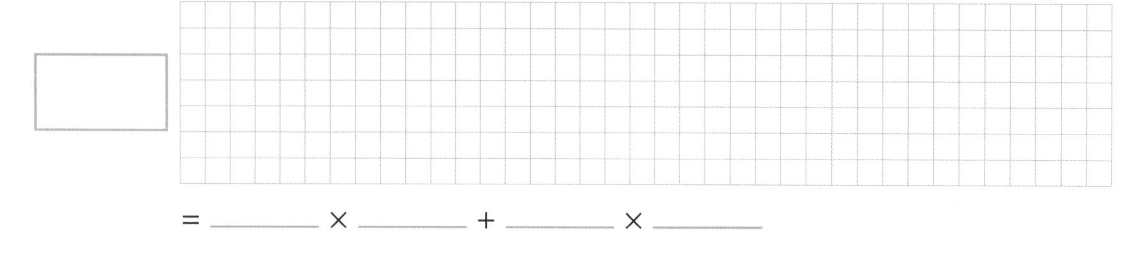

= ______ × ______ + ______ × ______

= ______ + ______

= ______

2 Solve the following problems. Remember to trade where necessary.

a

	3	3	7
×			5

b

	4	1	3
×			6

c

	2	8	4
×			8

d

	6	4	3
×			4

e

	2	9	3
×			3

f

	8	2	4
×			6

g

	2	3	8
×			9

h

	6	1	7
×			7

Challenge

1 Look for patterns in the answers to the following:

a

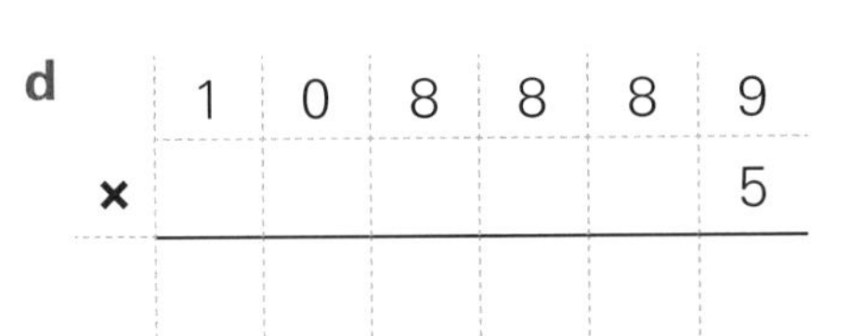

	2	6	3	8	9
×					8

b

	5	8	3	3	3
×					4

c

	5	3	3	9	1
×					7

d

	1	0	8	8	8	9
×						5

e

	1	0	5	5	5	6
×						6

f

	1	0	4	0	7	1
×						9

2 You will need to use some spare paper to solve the following problems.

a Alice buys six games at $32.65 each. How much does she pay altogether?

____

b Tom needs eight lengths of rope. Each piece needs to be 14.35 m. How much rope does he need altogether? ______________

c Jo's van can carry 240 kg in the back. She wants to buy some bags of soil. Each bag weighs 39.5 kg. How many bags can she take with her? ______________

d Audrey is putting panels on a wall. Each panel is 1320 mm wide. The wall is 12 m wide. How many boards should she get? ______________

3 Use a written strategy to solve the following problems.

a What is 15 × 23?

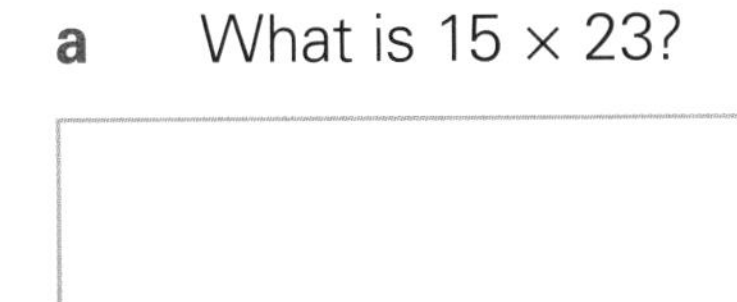

b What is 18 × 42?

c What is 14 × 63?

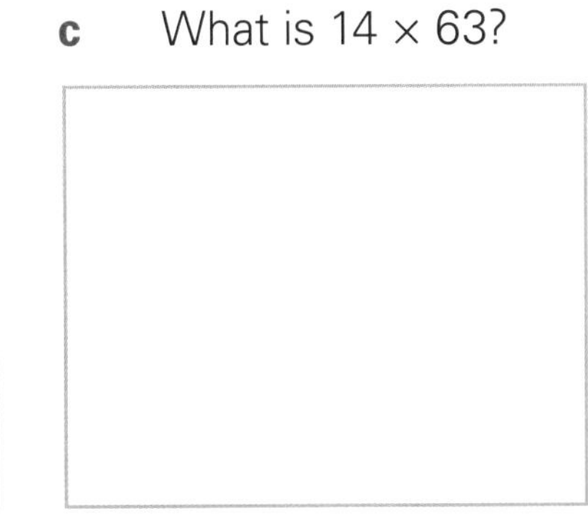

d What is 21 × 39?

e What is 42 × 36?

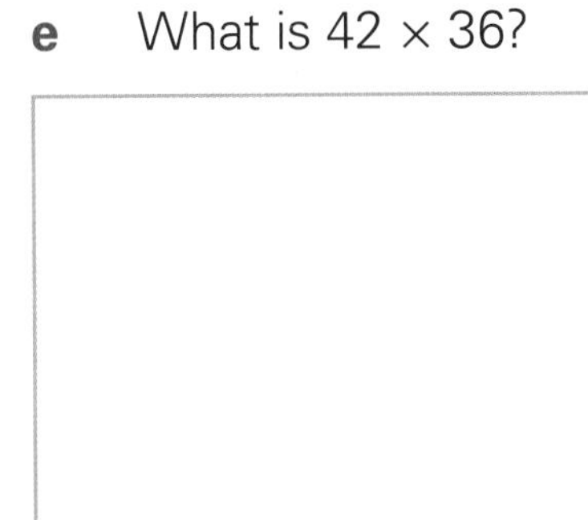

f What is 27 × 83?

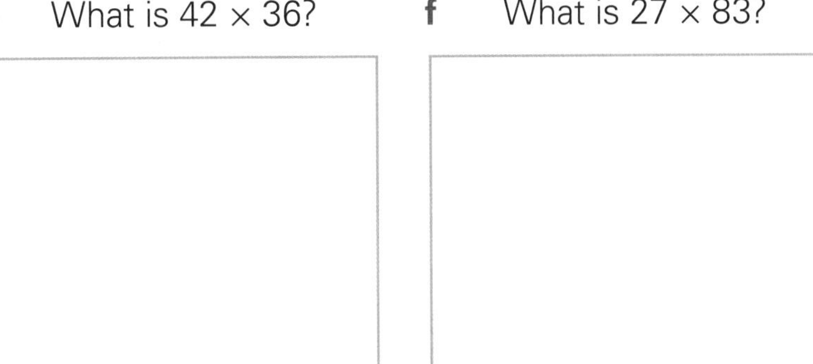

Mastery

1 We hear flying insects buzzing because of the sound their wings make when they are beating. The European midge is smaller than a mosquito, and its wings beat more than 1000 times per second!

Insect	Wing beats per second	Wing beats per minute	Wing beats per hour
Midge	1046		
Mosquito	595		
Fruit fly	315		
Wasp	247		
Fly	190		

a Use written strategies to calculate the number of wing beats per minute and per hour. Write your answers in the table.

b Time how long it takes you to write the word "multiplication" backwards (*noitacilpitlum*). How many times would a mosquito's wings beat in the time that it took you to write it? ______

c How many times would a fruit fly's wings beat if it flew for a full day? (You may wish to check your answer on a calculator.) ______

---

2 The Ferrari 458 is not the most expensive car in the world, but its price tag is over half a million dollars.

a If a Ferrari cost $526 950 and a car dealer sold one Ferrari a month, how much money would they receive in a year?

______

b With your teacher's permission, do some research about the most expensive cars in your country. Write some questions similar to the one about the Ferrari for others to answer. You could illustrate your questions. Make sure you know the answers!

Factors and multiples

Practice

1 For each row, circle the numbers that are **multiples** of the red number.

a	3	15	21	23	13	24	12	28	33	43	30	27	26
b	6	9	12	15	24	26	32	30	16	36	20	42	48
c	4	8	14	16	22	24	18	32	30	40	20	34	36
d	9	19	25	18	27	28	63	36	60	54	49	90	39
e	8	28	12	18	16	26	24	36	32	44	56	68	72

2 The **factors** of 6 are 1, 2, 3 and 6. Circle the factors of the red number in each row.

a	10	1	2	3	4	5	6	7	8	9	10		
b	16	1	2	3	4	5	6	7	8	9	10	14	16
c	12	1	2	3	4	5	6	7	8	9	10	11	12
d	20	1	2	3	4	5	6	7	8	9	10	15	20
e	27	1	2	3	4	5	6	7	8	9	10	21	27
f	18	1	2	3	4	5	6	8	9	10	12	18	20
g	22	1	2	3	4	5	6	8	9	10	11	20	22
h	15	1	2	3	4	5	6	8	9	10	12	15	20

3 Write the factors of each of the following numbers.

a 32 ☐ ☐ ☐ ☐ ☐ ☐

b 11 ☐ ☐

c 33 ☐ ☐ ☐ ☐

d 28 ☐ ☐ ☐ ☐ ☐ ☐

e 29 ☐ ☐

f 9 ☐ ☐ ☐

g 42 ☐ ☐ ☐ ☐ ☐ ☐

h 49 ☐ ☐ ☐

Do smaller numbers have fewer factors?

Challenge

1 Write the factors of each number below. Then write the common factors.

a The factors of 6 are: ____________

The factors of 10 are: ____________

The common factors of 6 and 10 are: ____________

b The factors of 20 are: ____________

The factors of 30 are: ____________

The common factors of 20 and 30 are: ____________

c The factors of 15 are: ____________

The factors of 25 are: ____________

The common factors of 15 and 25 are: ____________

d The factors of 21 are: ____________

The factors of 27 are: ____________

The common factors of 21 and 27 are: ____________

---

2 How do you know that:

a 992 is a multiple of 2? ____________

b 1221 is a multiple of 6? ____________

c 89 is **not** a multiple of 10? ____________

d 133 is **not** a multiple of 5? ____________

e 2324 is a multiple of 4? ____________

f 1006 is **not** a multiple of 6? ____________

---

3 What is the **lowest** common multiple of:

a 5 and 6? ________

b 4 and 6? ________

c 4 and 9? ________

d 3 and 8? ________

e 9 and 10? ________

f 7 and 11? ________

---

4 Find a common multiple of 3 and 6 between 30 and 40. ____________

---

5 Find a common multiple of 3 and 8 between 70 and 80. ____________

Mastery

1 The number 12 is special because it is the first counting number that has six factors.

a Write the factors of 12. ______________________

b If you add the factors (**apart from 12 itself**) you will see that the total is greater than 12. What is the total of the factors (apart from 12)? __________

Abundant numbers

2 The number 12 is called an **abundant number**. This is because the total of its factors (apart from the number itself) is greater than 12.

a One of the numbers 16 and 18 is also an abundant number. Carry out the test from question 1 to find out which.

____

____

b Apart from the abundant numbers you have already identified, there are another 20 abundant numbers on a hundred grid. Find as many of them as you can. As you find one, highlight or circle it on the hundred grid. Be ready to explain how you are sure each number is an abundant number. Something to think about before you start: do you think any odd numbers will be abundant numbers?

1	2	3	4	5	6	7	8	9	10
11	12	13	14	15	16	17	18	19	20
21	22	23	24	25	26	27	28	29	30
31	32	33	34	35	36	37	38	39	40
41	42	43	44	45	46	47	48	49	50
51	52	53	54	55	56	57	58	59	60
61	62	63	64	65	66	67	68	69	70
71	72	73	74	75	76	77	78	79	80
81	82	83	84	85	86	87	88	89	90
91	92	93	94	95	96	97	98	99	100

UNIT 1: TOPIC 9
Divisibility

Practice

1 a Circle the numbers that are exactly divisible by 5.

52 45 70 58 310 554 990 2995 4551 5054 4440 7110

b Circle the numbers that are exactly divisible by 6.

18 26 38 72 810 315 603 4314 4126 3054 2226 2626

2 All the multiples of 8 are even numbers, but not every even number is divisible by 8.

a Write a two-digit even number that is **not** divisible by 8. ________

b Circle the even numbers that are exactly divisible by 8.

8 14 28 32 38 72 78 148 168 880 1148

2168 2356 1248 3880 7832 3486 3488 5728 4108

3 Without using any numbers already on this page, write:

a a four-digit number that is exactly divisible by 5. ________

b a three-digit number that is exactly divisible by 8. ________

c a four-digit number that is exactly divisible by 6. ________

d a four-digit number that is exactly divisible by 8. ________

4 How do you know that 351 is divisible by 3?

____

Challenge

1. a How do you know that 43 is a prime number?

   ______

   b Name a prime number other than 43 between 40 and 50. ______

2. Look at this row of numbers:

51	52	53	54	55	56	57	58	59	60

Write the numbers that are exactly divisible by:

a 3 ______ b 5 ______

c 8 ______ d 9 ______

e 3 and 6 ______ f 2 and 4 ______

g 3, 6 and 9 ______ h 2, 4 and 8 ______

3. Now, look at this row of numbers:

71	72	73	74	75	76	77	78	79	80

a Which is the only number in the row that is divisible by 2, 3, 4, 6, 8 and 9?

______

b The answer to question 3a has many factors. What are all its factors?

______

c Which are the prime numbers in the row of numbers? ______

4. a There are five two-digit numbers that each have 12 factors. Write the numbers and make a list of the factors for each number.

   b What are the common factors of the five numbers you found in question 4a?

   ______

Mastery

1 On the previous page you found the common factors of five numbers: 60, 72, 84, 90 and 96. You found that all the numbers were divisible by 2, 3 and 6. (Since every number is divisible by 1, we won't worry about 1 in this task.)

60 is the lowest number in the list that is divisible by all three numbers. So, we can say that 60 is a common multiple of 2, 3 and 6. However, 60 is not the **lowest** common multiple of 2, 3 and 6.

Let's find the lowest common multiple of just two numbers: 2 and 3. We start by listing the multiples of each number and stop when we find a number that appears in each list.

- Multiples of **2**: 2, 4, (6)
- Multiples of **3**: 3, (6)

We can stop because we have already found the first (lowest) common multiple: 6

Use this method to find the lowest common multiple of the following:

a 3 and 4

- Multiples of **3**: ____________
- Multiples of **4**: ____________

The lowest common multiple of 3 and 4 is: ________

b 3 and 8

- Multiples of **3**: ____________
- Multiples of **8**: ____________

The lowest common multiple of 3 and 8 is: ________

---

2 Use the same method to find the lowest common multiple of three numbers:

a 3, 4 and 8

Multiples of **3**: ____________

Multiples of **4**: ____________

Multiples of **8**: ____________

The lowest common multiple of 3, 4 and 8 is: ____________

b 6, 15 and 20

Multiples of **6**: ____________

Multiples of **15**: ____________

Multiples of **20**: ____________

The lowest common multiple of 6, 15 and 20 is: ____________

---

3 On a building site, Chet can carry 3 bricks at a time, Lucy can carry 5 bricks at a time and Hercules can carry 8 bricks at a time. What is the lowest number of bricks they must have carried if they each carried the same number over the morning? ________

UNIT 1: TOPIC 10
Division written strategies

Practice

1 Split the numbers to solve the following division problems.

a What is 142 ÷ 2?

142 ÷ 2 is the same as
_____ ÷ 2 and 42 ÷ 2

_____ ÷ 2 = _____

42 ÷ 2 = _____

So, 142 ÷ 2 = _____ + _____
= _____

b What is 128 ÷ 4?

128 ÷ 4 is the same as
_____ ÷ 4 and _____ ÷ 4

_____ ÷ 4 = _____

_____ ÷ 4 = _____

So, 128 ÷ 4 = _____ + _____
= _____

c What is 135 ÷ 5?

135 ÷ 5 is the same as
_____ ÷ 5 and _____ ÷ 5

_____ ÷ 5 = _____

_____ ÷ 5 = _____

So, 135 ÷ 5 = _____ + _____
= _____

d What is 324 ÷ 3?

324 ÷ 3 is the same as
_____ ÷ 3 and _____ ÷ 3

_____ ÷ 3 = _____

_____ ÷ 3 = _____

So, 324 ÷ 3 = _____ + _____
= _____

2 Solve the following division problems. There is no remainder.

a $2\overline{)86}$ b $4\overline{)88}$ c $3\overline{)96}$ d $2\overline{)68}$

e $4\overline{)864}$ f $5\overline{)755}$ g $7\overline{)847}$ h $8\overline{)976}$

i $4\overline{)6536}$ j $5\overline{)5730}$ k $7\overline{)7371}$ l $6\overline{)8352}$

m $5\overline{)63\,480}$ n $6\overline{)69\,246}$ o $3\overline{)43\,521}$

Challenge

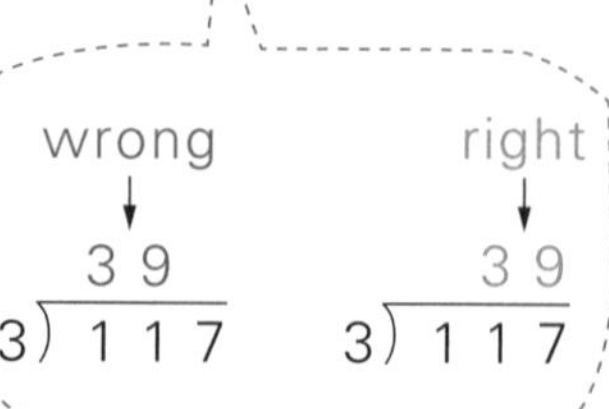

1 Solve these division problems.

a 8) 5 6 **b** 9) 7 2 **c** 5) 4 5

d 6) 5 4 **e** 8) 2 7 2 **f** 6) 3 8 4

g 7) 2 7 3 **h** 4) 3 9 6 **i** 4) 1 5 7 2

j 7) 5 7 3 3 **k** 3) 2 5 4 1 **l** 8) 5 2 3 2

2 Use "r" to show the remainder for the following:

a 8) 9 9 **b** 6) 9 3 **c** 9) 9 2

d 7) 9 5 **e** 5) 7 3 7 **f** 3) 9 2 5

g 7) 8 3 6 **h** 9) 9 2 3 **i** 7) 9 3 5 4

j 5) 3 5 7 4 **k** 6) 8 3 4 8 **l** 4) 7 4 9 5

m 3) 8 4 8 9 6 **n** 5) 1 6 1 9 3 **o** 8) 4 9 1 9 1

Mastery

1 All the following tasks use the same numbers. Make sure you deal with the remainder appropriately each time, and be ready to show your working out.

a A company manufactures 23 255 marbles. They are shared between 6 shops. How many marbles will each shop receive? ______________

b A store orders 23 255 cups. Each carton can hold no more than six cups. What is the smallest number of cartons that is needed? ______________

c Six people share \$23 255. How much does each person get? ______________

---

2 Airports can be very busy places. This table shows approximately how many people use some of the busiest airports in the world each week.

Airport	Number of passengers per week
Beijing, China	1 729 615
Chicago, USA	1 479 615
Dubai, United Arab Emirates	1 500 192
Hartsfield–Jackson, Georgia, USA	1 951 730
London Heathrow, UK	1 442 115
Paris CDG, France	1 264 807
Sydney, Australia	762 615
Tokyo, Japan	1 448 461

Airport	Number of passengers per day (rounded)

a The list of airports in the table above is in alphabetical order. Reorder the list in the second table above, starting with the busiest airport and ending with the least busy.

b Use spare paper to calculate the number of passengers each airport has per day. Round each answer to the nearest whole number, and be ready to show your working out.

UNIT 2: TOPIC 1
Comparing and ordering fractions

Practice

1 Shade the shapes as specified below.

a Shade one third. b Shade $\frac{5}{6}$. c Shade half. d Shade $\frac{3}{5}$. e Shade $\frac{3}{8}$.

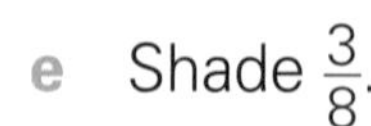

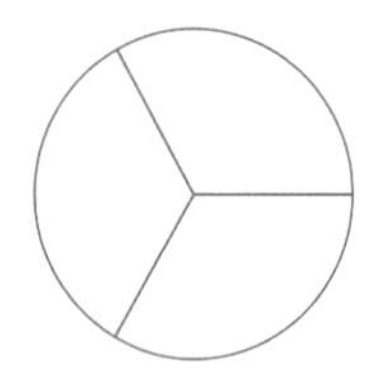 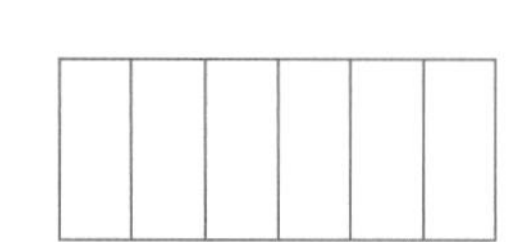 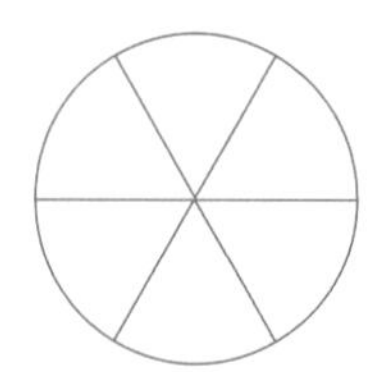 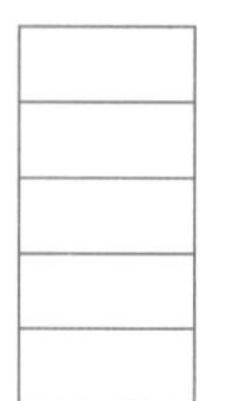 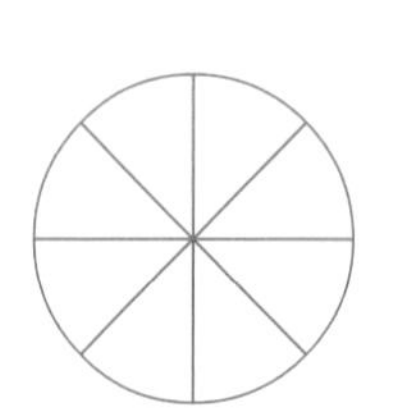

2 Shade each group to match its fraction.

a $\frac{7}{10}$ b $\frac{3}{4}$ c $\frac{2}{3}$ d $\frac{1}{3}$

3 What fraction of each group is shaded?

a 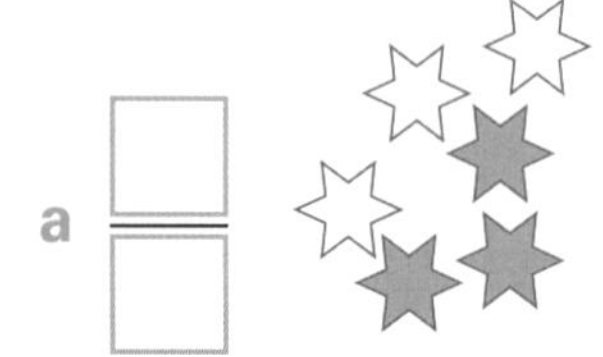 b 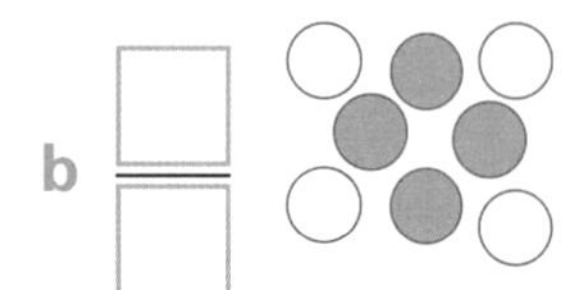 c 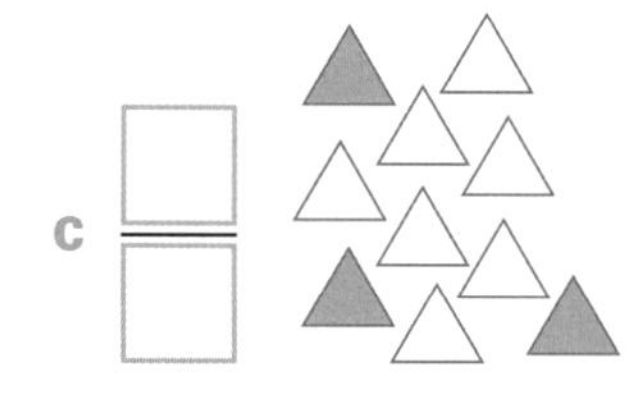 d 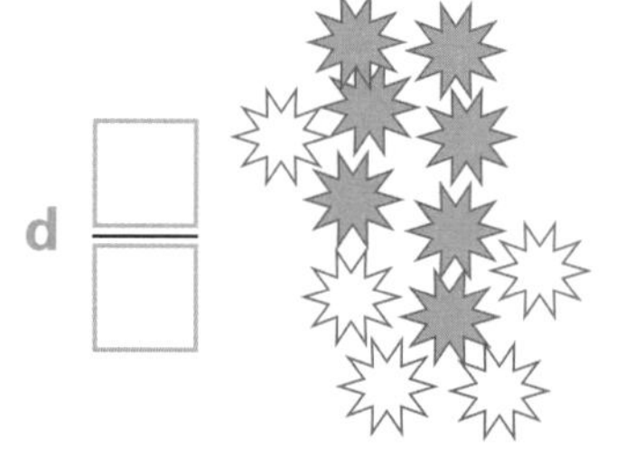

4 Follow the instructions to shade the circles. Complete each sentence to show which fraction is bigger.

a Shade $\frac{1}{3}$ and $\frac{1}{4}$.

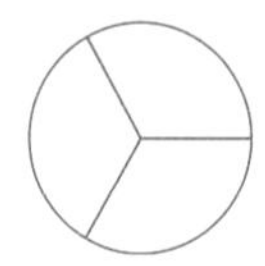 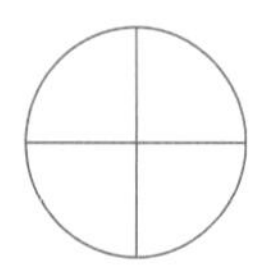

The bigger fraction is: ____

b Shade $\frac{1}{5}$ and $\frac{1}{6}$.

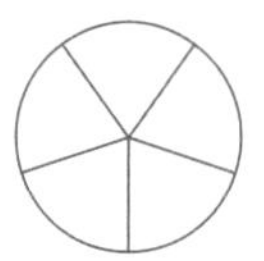 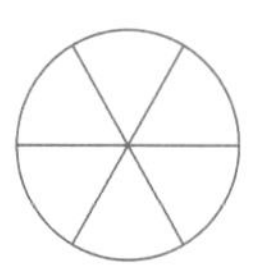

The bigger fraction is: ____

c Shade $\frac{1}{8}$ and $\frac{1}{6}$.

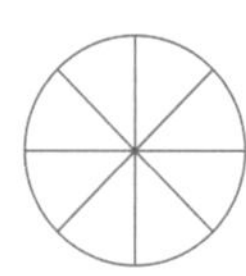 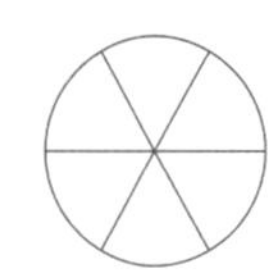

The bigger fraction is: ____

d Shade $\frac{1}{9}$ and $\frac{1}{10}$.

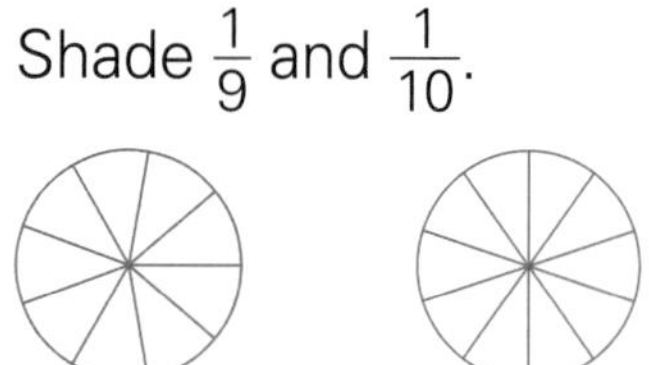

The bigger fraction is: ____

Challenge

1 Fill in the fractions on each number line.

a

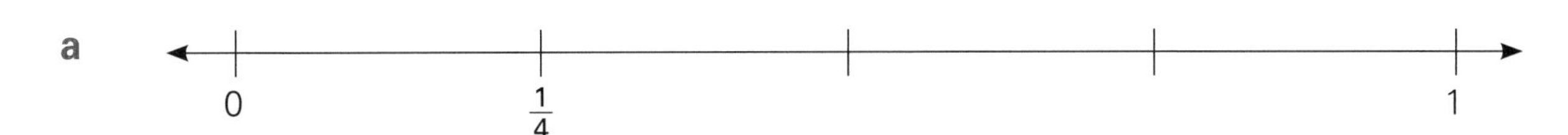

b

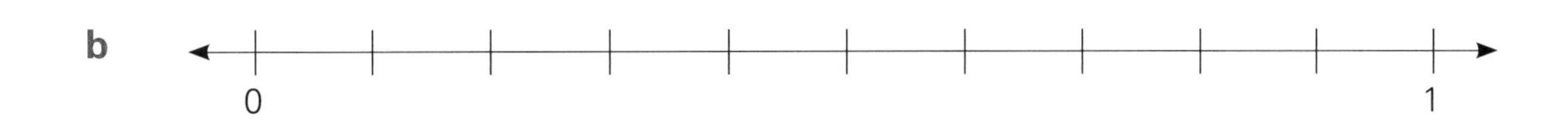

c

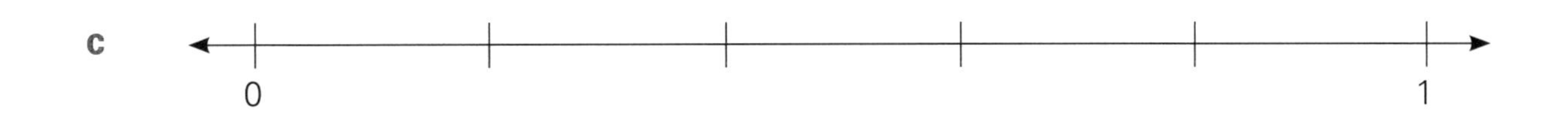

d

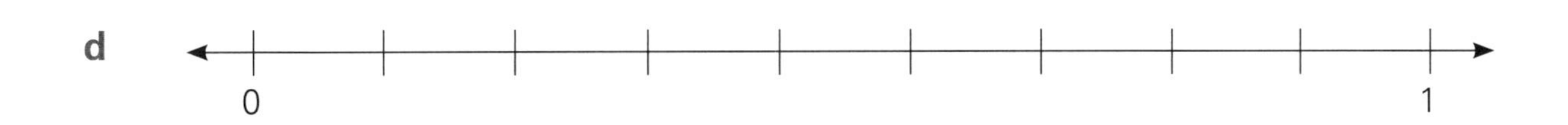

e

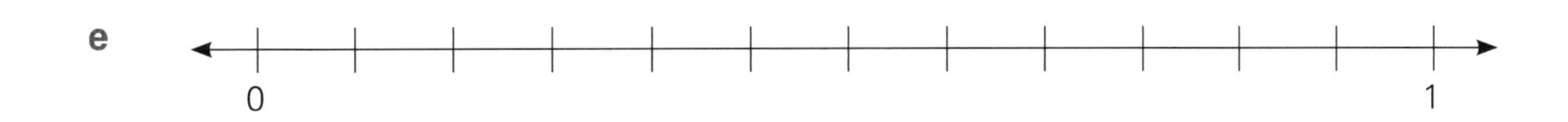

---

2 Which is the bigger fraction in each pair below? Use the number lines in question 1 to help you.

a $\frac{7}{10}$ or $\frac{3}{5}$? ________ **b** $\frac{2}{10}$ or $\frac{1}{4}$? ________ **c** $\frac{3}{5}$ or $\frac{1}{2}$? ________

d $\frac{1}{9}$ or $\frac{1}{12}$? ________ **e** $\frac{9}{10}$ or $\frac{8}{9}$? ________ **f** $\frac{3}{4}$ or $\frac{8}{10}$? ________

---

3 Order the groups of fractions from smallest to largest.

a $\frac{6}{10}$, $\frac{3}{10}$, $\frac{10}{10}$, $\frac{2}{5}$, $\frac{1}{2}$, $\frac{1}{10}$, $\frac{1}{5}$ ________________________________

b $\frac{1}{10}$, $\frac{4}{12}$, 1, $\frac{1}{12}$, $\frac{1}{4}$, $\frac{4}{5}$, $\frac{4}{10}$ ________________________________

c 1, $\frac{1}{9}$, $\frac{1}{10}$, $\frac{1}{5}$, $\frac{1}{2}$, $\frac{1}{12}$, $\frac{1}{4}$ ________________________________

Mastery

1 In the student book you looked at how difficult it is to fold a piece of paper in half more than eight times. You also learned that the bigger the number in the denominator, the smaller each part becomes.

In this task you are going to cut, not fold, paper in half. This challenge is a practical one and will test your powers of accuracy and patience.

Start with two pieces of coloured paper that are exactly the same size (for example, 10 cm × 15 cm).

- Glue one of the pieces onto a page and label it "one whole".
- Fold the other in half, cut out **one** half, glue it onto the page and label it "one half".
- Fold and cut the remaining piece in half again.
- Work out the fraction, glue it onto the page and label it.

one whole
one half
one quarter
one eighth

How small can you go?

2 Sean's local cake shop sells cakes by the slice and will cut their cakes into any number of pieces for customers. If Sean bought between 4 and 5 cakes, suggest what fraction of cakes he might have bought in total as a mixed number and how many people he was feeding. Show three different solutions.

3 Sean needed to buy 25 slices of cake. Suggest the total amount of cake he might have bought as an improper fraction and a mixed number for three different unit fractions. For example if he bought 25 quarter slices, he would have $2\frac{5}{4}$ or $6\frac{1}{4}$ cakes.

UNIT 2: TOPIC 2
Adding and subtracting fractions

Practice

1 Write the number sentence.

a 

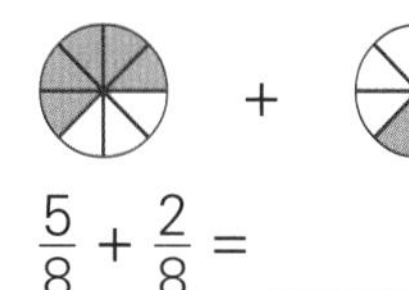

$\frac{5}{8} + \frac{2}{8} =$ ______

b 

____

c 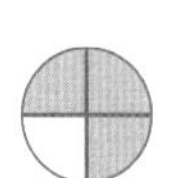 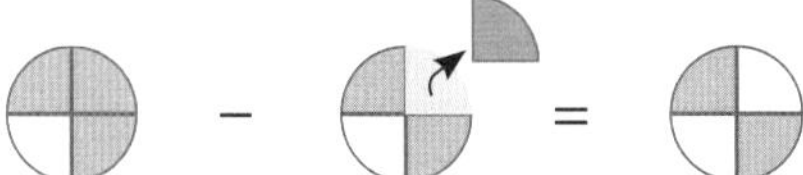

____

d  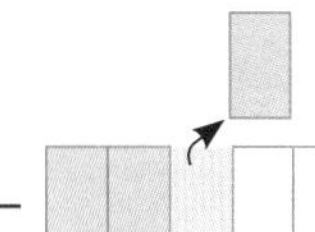

____

---

2 Write the number sentence. Shade the third shape to show the answer.

a 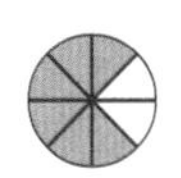 + =

____

b 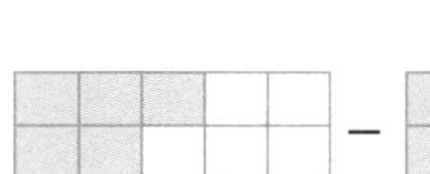 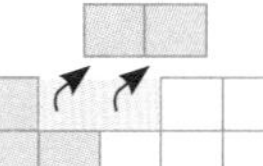

____

c 

____

d  

____

e 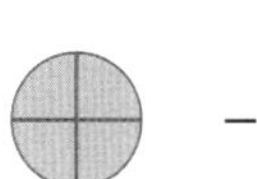 

____

f 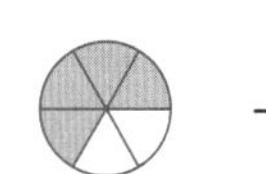 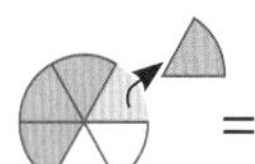 

____

---

3 Complete the number sentences. Use two colours to shade each diagram so that it matches its number sentence.

a $\frac{1}{5} + \frac{2}{5} =$ ☐ 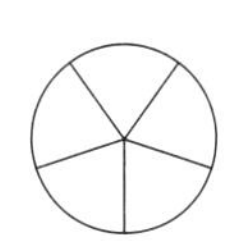

b $\frac{3}{6} + \frac{2}{6} =$ ☐

c $\frac{2}{9} + \frac{5}{9} =$ ☐

d $\frac{1}{8} + \frac{3}{8} =$ ☐

---

4 Complete the subtraction sentences. The diagrams may help you.

a $\frac{7}{10} - \frac{3}{10} =$ ☐

b $\frac{6}{7} - \frac{3}{7} =$ ☐

c $\frac{4}{5} - \frac{1}{5} =$ ☐ 

d $\frac{2}{3} - \frac{2}{3} =$ ☐ 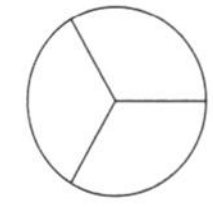

Challenge

1 Shade the shapes to show the answer to each addition. Write the answer as an improper fraction and as a mixed number.

a 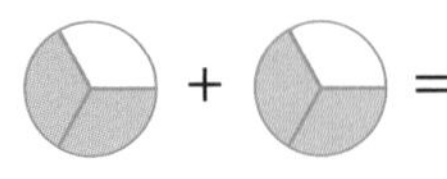 

____

b 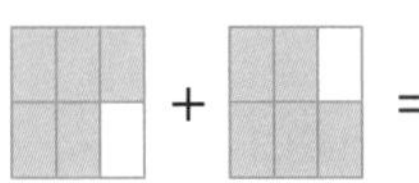 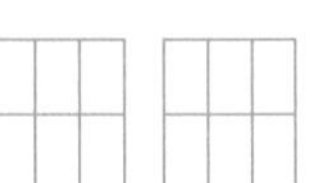

____

c 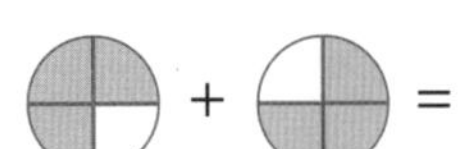 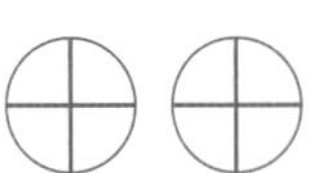

____

d 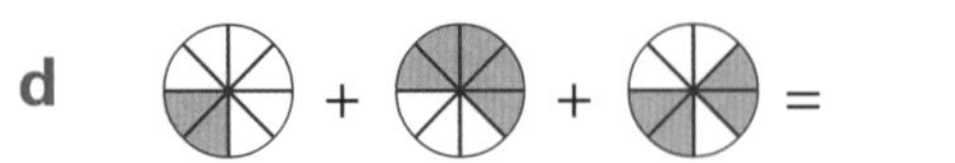 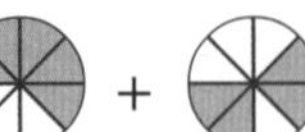 

____

---

2 Use the number lines to help you add and subtract. Use improper fractions and mixed numbers where appropriate.

a $\frac{1}{4} + \frac{3}{4} =$ ______________________

b $2 - 1\frac{3}{8} =$ ______________________

c $\frac{7}{10} + \frac{7}{10} =$ ______________________

d $\frac{5}{6} + \frac{4}{6} =$ ______________________

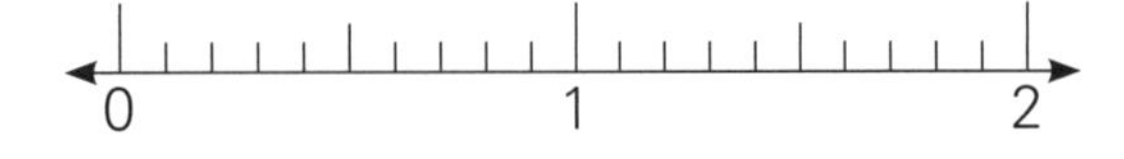

e $\frac{4}{5} + \frac{3}{5} + \frac{4}{5} =$ ______________________

f $2\frac{1}{2} - 1\frac{1}{4} =$ ______________________

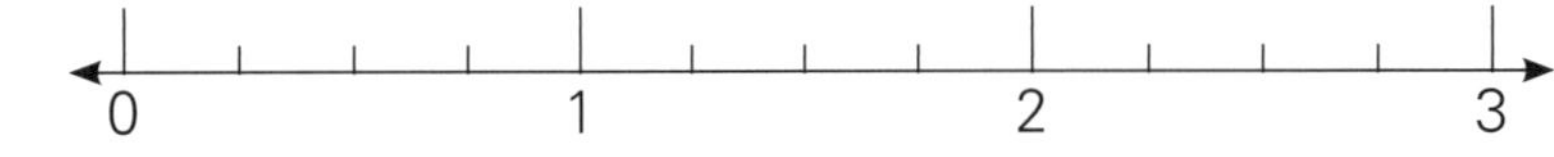

g $\frac{9}{10} + 1\frac{3}{10} + \frac{7}{10} + \frac{3}{10} =$ ______________________

Mastery

1 Write three different number sentences. They all have the same answer.

a ____ + ____ $= 1\frac{1}{2}$ b ____ + ____ $= 1\frac{1}{2}$ c ____ + ____ $= 1\frac{1}{2}$

2 Fill the gaps to make these statements true:

a $\frac{\square}{8} + \frac{\square}{8} > 1$ b $1\frac{\square}{10} - \frac{\square}{10} < 1\frac{3}{10}$

c $\frac{\square}{4} + \frac{\square}{2} < 1$ d $\frac{\square}{\square} + \frac{\square}{\square} = \frac{1}{5}$

3 Write two fractions that have a difference of $\frac{3}{4}$. ____________________

4 Three children share two pizzas. No one eats the same amount as anybody else. No one eats a whole number of pizzas. Their names are in order from the person who eats the least to the person who eats the most. Write the amount that each person could eat. Make sure the total is two pizzas.

Tran: ________ Jim: ________ Sam: ________

Total = 2 pizzas

5 A birthday cake is easy to divide up if it is cut into quarters. Imagine you are a teacher and someone brings a cake to share between 24 students. The first cut would be in half – and then what? What would the next cuts be? Draw the cake on one of these circles.

If you find that task easy, decide what cuts would be needed to divide a cake into equal slices for 27, 28 and 30 people. Why would sharing between 29 people be even harder? How would you do that?

UNIT 2: TOPIC 3
Decimal fractions

Practice

1 Write the shaded part of each hundred grid as a fraction and as a decimal.

a 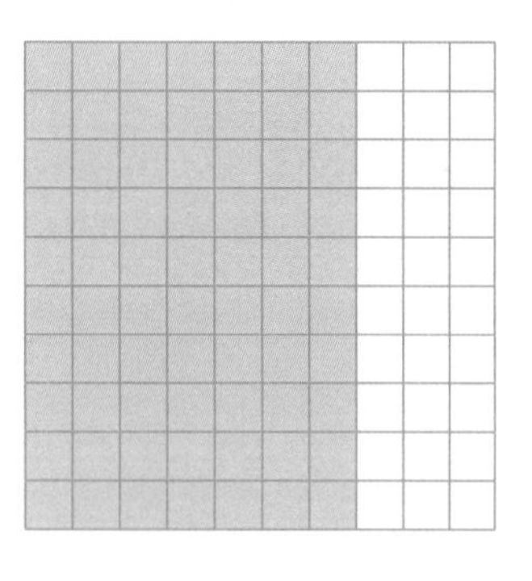

b 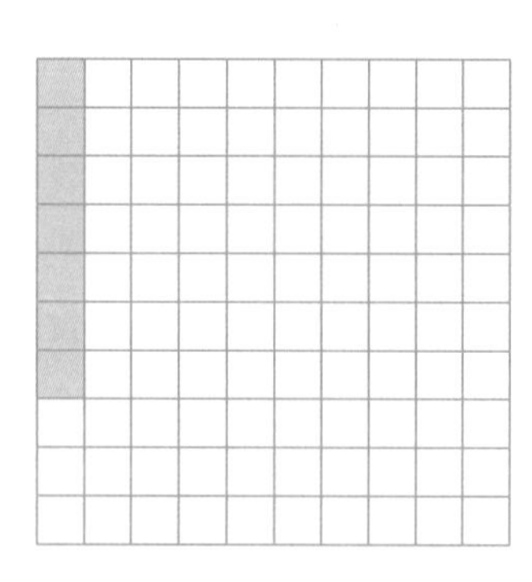

c 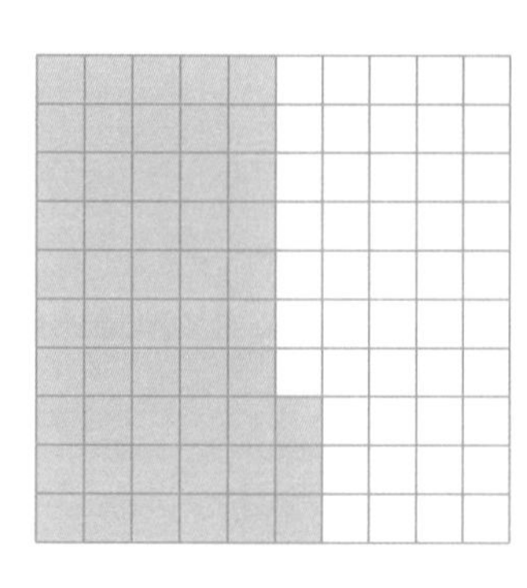

2 Shade the diagrams to match the decimals. Write the fraction for each.

a 0.2 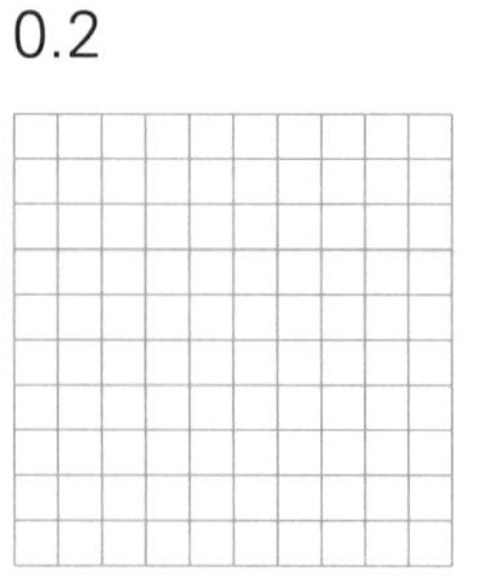

b 0.17 

c 0.01 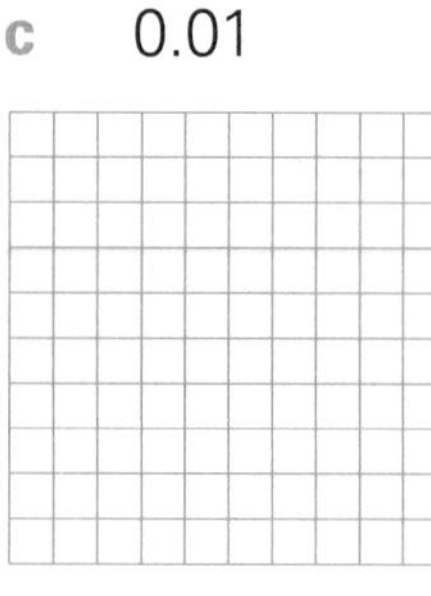

d 0.69 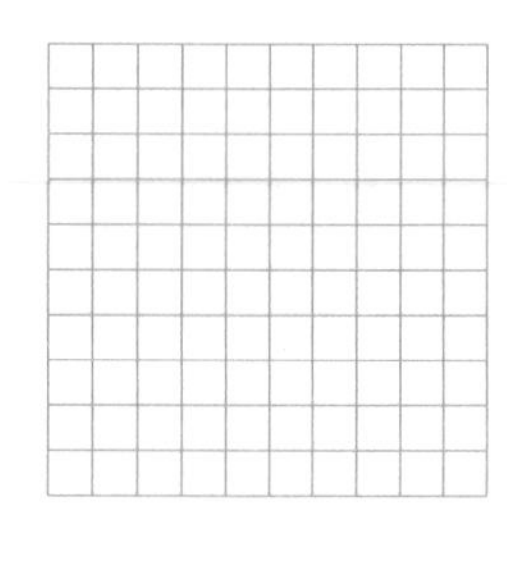

3 Write the shaded part as a decimal and as a fraction.

a 0. ____ $\frac{\square}{1000}$

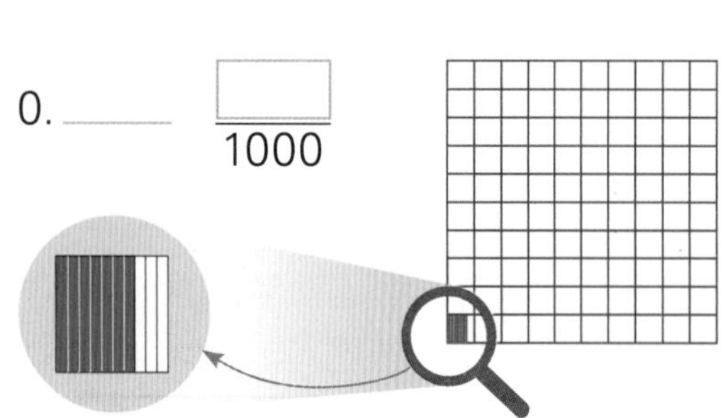

b 0. ____ $\frac{\square}{1000}$

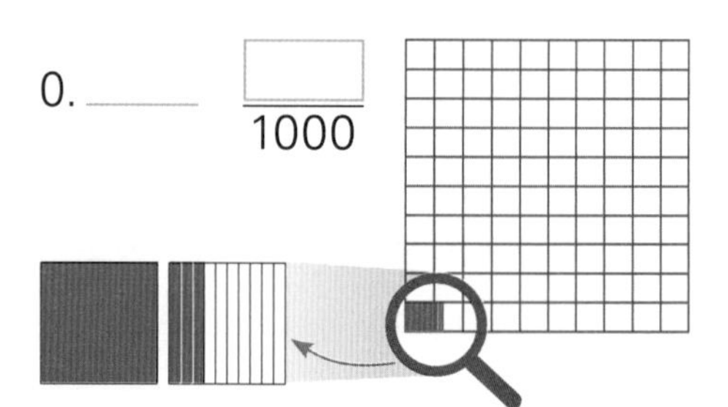

4 Write as decimals:

a $\frac{115}{1000}$ ____ b $\frac{87}{100}$ ____ c $\frac{17}{1000}$ ____

5 Write as fractions:

a 0.003 ____ b 0.45 ____ c 0.862 ____

Challenge

1 Complete the following by using the symbol **>**, **<** or **=**.

a 0.001 ______ 0.01　　**b** $\frac{5}{100}$ ______ 0.005　　**c** $\frac{350}{1000}$ ______ 0.35

d 0.01 ______ 0.009　　**e** $\frac{25}{1000}$ ______ 0.25　　**f** $\frac{10}{1000}$ ______ 0.01

g 0.99 ______ $\frac{99}{1000}$　　**h** $1\frac{5}{10}$ ______ 1.05　　**i** $\frac{199}{1000}$ ______ 0.2

---

2 Look carefully at the start and end points on the following lines. Write the correct decimals in the boxes.

a 0 … 1

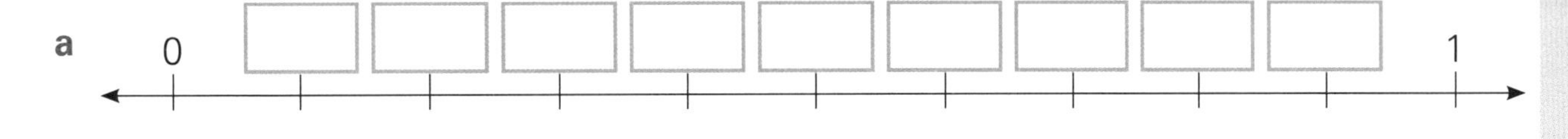

b 3 … 4

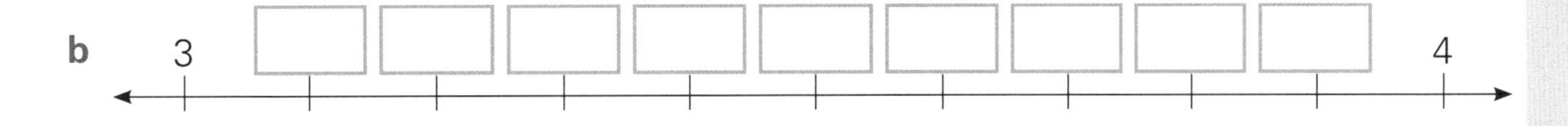

c 0.1 … 0.2

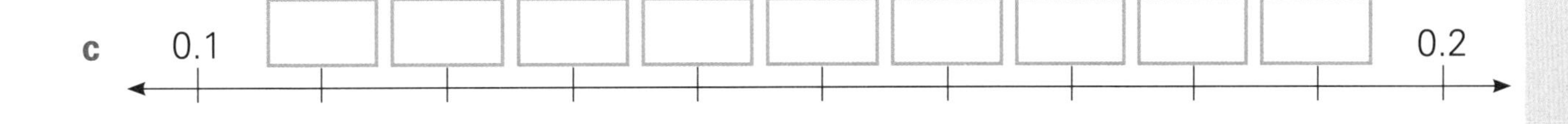

d 0.03 … 0.04

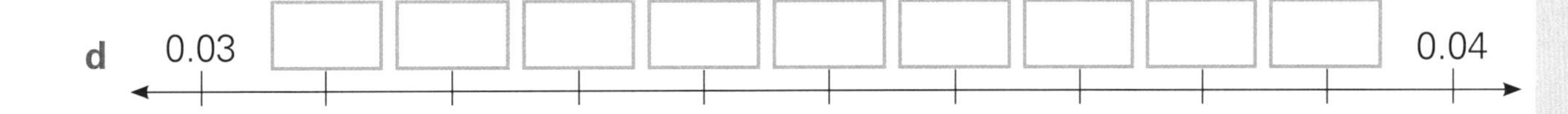

---

3 Put the fractions and decimals in order from **smallest** to **largest**.

a 0.8, 0.3, 0.4, $\frac{9}{10}$, $\frac{7}{10}$

____

b $\frac{2}{100}$, 0.07, $\frac{3}{10}$, 0.05, 0.01

____

c 0.8, 0.008, 0.08, $\frac{10}{1000}$, 0.1

____

d 0.5, 0.05, $\frac{5}{1000}$, $\frac{15}{1000}$, 0.051

____

e 0.4, $\frac{4}{100}$, $\frac{404}{1000}$, 4.004, 0.444

____

f 2.5, 0.25, 0.025, 2.55, 0.255

____

Mastery

1 If you divide 3 by 7 on a calculator, you see a lot of decimal places!

0.428571428571429

You know the value of the 4, the 2 and the 8, but what about the rest? In this task you are going to try to find out.

In Unit 2, Topic 1, you had to recognise fractions after cutting a piece of paper in half time and time again. This task is similar, but it involves decimals.

- You need two squares of paper that are each 10 cm × 10 cm.
- Divide one of the whole squares into tenths and cut out $\frac{1}{10}$ to see what 0.1 looks like.
- Cut out another tenth and divide it into ten equal parts to see what 0.01 looks like.
- Next, divide a hundredth into ten equal parts to see what 0.001 looks like.
- Next … well, you get the idea! How small can you get the pieces? Do you know their names?
- Glue your whole square and the decimal fractions onto paper or card and label them.

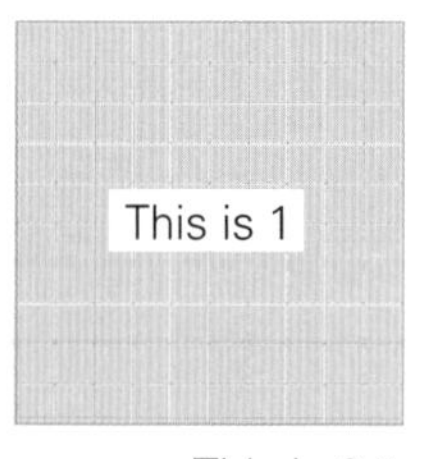

This is 0.1

This is 0.01

This is 0.001

If you want to make decimal fractions but you find the pieces of paper are too small, start with a bigger whole square, say 20 cm × 20 cm.

2 The number of decimal places that you see when you divide 3 by 7 on a calculator depends on the way the calculator has been programmed. In fact, the first six decimal places keep being repeated and repeated. How many of the decimal digits can you find the value of?

UNIT 2: TOPIC 4
Percentages

Practice

1 Write the shaded part as a fraction, as a decimal and as a percentage.

a

Fraction: ______

Decimal: ______

Percentage: ______

b

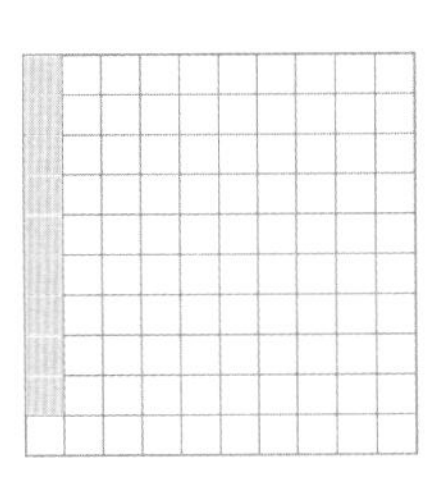

Fraction: ______

Decimal: ______

Percentage: ______

c

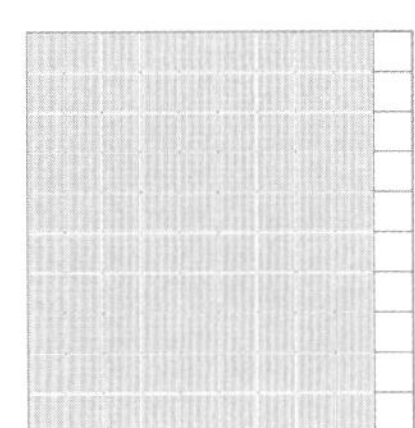

Fraction: ______

Decimal: ______

Percentage: ______

2 Shade the grid. Fill in the gaps.

a

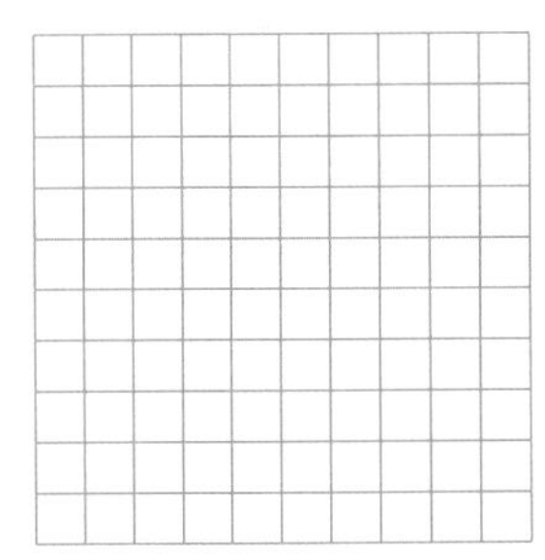

Fraction: ______

Decimal: 0.45

Percentage: ______

b

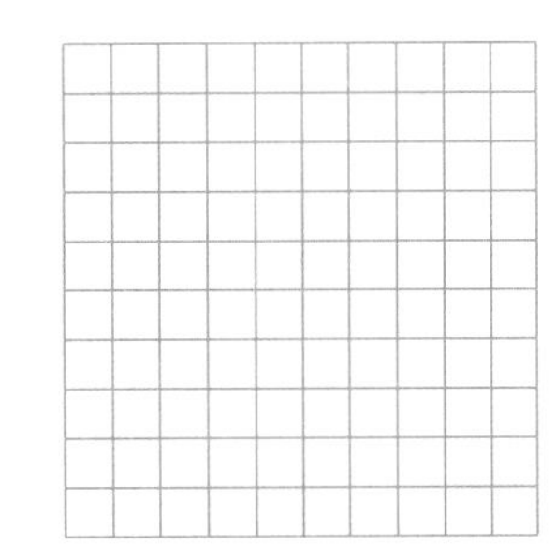

Fraction: $\frac{4}{10}$

Decimal: ______

Percentage: ______

c

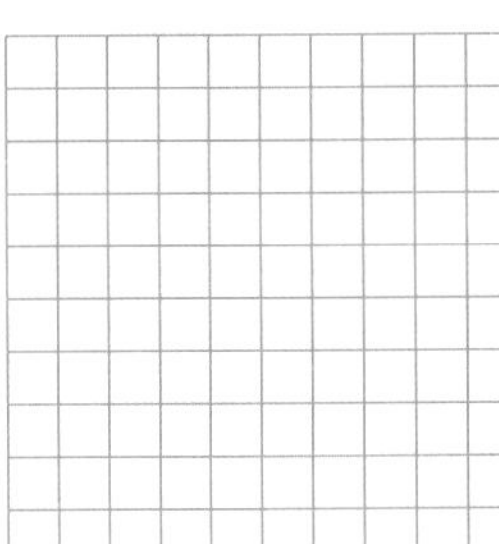

Fraction: ______

Decimal: ______

Percentage: 34%

d

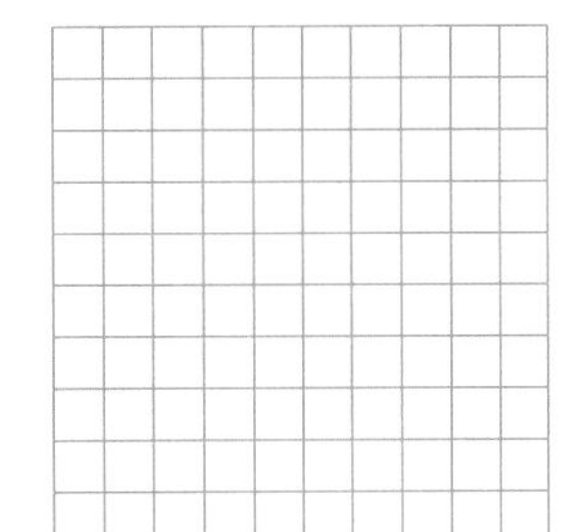

Fraction: $\frac{5}{100}$

Decimal: ______

Percentage: ______

3 Complete the following table:

	Fraction	Decimal	Percentage
a		0.15	
b	$\frac{1}{10}$		
c			85%
d	$\frac{5}{100}$		
e			39%
f	$\frac{7}{100}$		

4 Write **>**, **<** or **=** in the gaps below.

a 35% ______ 0.35

b 0.19 ______ 20%

c $\frac{5}{10}$ ______ 0.05

d 90% ______ $\frac{9}{10}$

e 0.7 ______ 68%

Challenge

1 Order the fractions, decimals and percentages in each row from **smallest** to **largest**.

a 0.4 4% $\frac{14}{100}$ 0.045

____

b 73% $\frac{3}{4}$ 0.72 74%

____

c $\frac{3}{10}$ 33% $\frac{13}{100}$ 0.03

____

d 9.5 95% 0.095 9.05

____

e $\frac{15}{100}$ 1.5 0.015 5%

____

f $\frac{23}{100}$ 0.024 $\frac{1}{4}$ 2.5%

____

2 There are 40 stars in the space here.

a Follow these instructions to colour the stars:

- Colour 10% red.
- Colour 25% blue.
- Colour $\frac{1}{4}$ green.
- Colour 20% yellow.

b How many stars are left over? Write the number as a fraction, as a decimal and as a percentage.

____

3 This time there are 50 stars.

a Follow these instructions to colour the stars:

- Colour 100% of the stars.
- Colour more than 50% blue.
- Colour 20% red.
- Colour more red than green.
- Colour the same percentage green and yellow.

b Write the number and percentage for each colour.

____

Mastery

1 It is not unusual to see this sort of headline in the media:

Daily Blah-Blah

SPORTS LATEST:

New coach demands 110% effort from team!

How would you explain that it is not possible to give 110% effort if you were on the team?

---

2 People selling things often use percentages in their advertisements. These signs use decimals or fractions instead. Rewrite each sign with the appropriate percentage.

a Today only: $\frac{1}{10}$ off everything

b This week: $\frac{1}{4}$ discount

c Save 0.15 immediately!

d Contains $\frac{2}{10}$ less sugar!

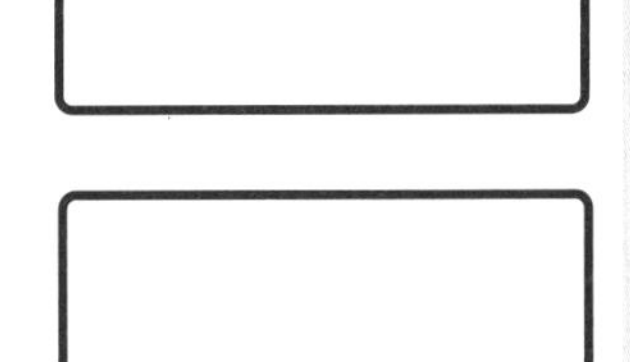

e Up to $\frac{70}{100}$ off STOREWIDE!

f Buy two and save 0.05

---

3 Stores often advertise a sale with signs that say something like "Up to 75% off".

a Imagine you are shopping with a friend. Your friend sees the "Up to 75% off" sign and says, "Wow! We can save 75% on everything!" How could you explain that the sign does not mean what your friend thinks it does?

b Why do you think shops display signs like "Up to 75% off"?

c Look in newspapers and magazines for advertisements offering "up to …" discounts. You could attach them to this page.

Financial plans

Practice

1 A class of 25 students needs new sports equipment. The teacher asks the students to help choose what to buy. They have $500 to spend. If the money is divided equally between them, how much will each student have to spend? ________

---

2 The teacher shows the class a sports catalogue. They can choose what they want to buy from the catalogue.

Remember: Round to the nearest 5c or 10c.

Small balls: $0.59 each

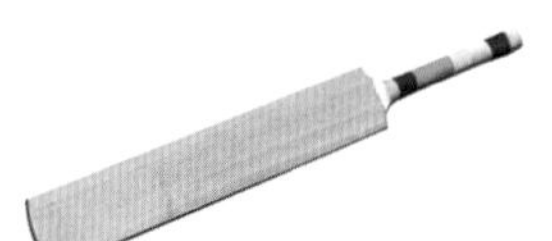

Bats: $7.99 each

Hoops: $2.49 each

Skipping ropes: $1.99 each

Pack of five balls: $14.99

Using rounding where necessary, how much would it cost for:

a two tennis balls? ________

b five skipping ropes? ________

c three bats? ________

d five hoops? ________

e two packs of five balls? ________

f a bat and three small balls? ________

---

3 a In the pack of five balls, what is the average (mean) price of each ball?

____

b Each ball in the pack has a different value. Using rounded prices, what is your estimate for the value of each ball?

____

---

4 a The hoops can be bought in a pack of six for $11.99. How much can be saved by buying six in a pack? ________

b The skipping ropes are sold in a pack of 20 at a price of $1.80 per rope. How much money can be saved by buying 20 skipping ropes in a pack compared with 20 individual skipping ropes? ________

OXFORD UNIVERSITY PRESS

Challenge

1 The teacher notices that there will be a sale soon at the sports store. Look at the percentage discounts for each item. Write the amount of discount and the new price. Round the prices where necessary.

a

Bat:
10% discount
Amount of discount:

______

New price:

______

b

Pack of balls:
10% discount
Amount of discount:

______

New price:

______

c

Single skipping ropes:
25% discount
Amount of discount:

______

New price:

______

d

Small balls:
half price
Amount of discount:

______

New price:

______

e

Hoops:
10% discount
Amount of discount:

______

New price:

______

f

Pack of 20 ropes:
10% discount
Amount of discount:

______

New price:

______

---

2 **a** Using the original catalogue price from page 44, how much would it cost in total to buy a hoop, a bat, a rope, a pack of balls and four small balls for each child in the class?

______

b How much more than the budget of $500 would that cost? ______

---

3 The teacher says that they do not need a pack of five balls for each child. They decide to get only six packs.

a What is the new total? ______

b What is the difference between the new total and the class budget?

______

Mastery

1 Imagine you are in the class that has the \$500 budget for sports equipment. Decide how you would spend the money so that it was fair for the whole class.

For your calculations assume that the sale has started and that the special offers for buying in bulk can be used. You will probably need spare paper for this activity.

Begin by rewriting the prices, including the discounts.

- Small ball: ____________
- Pack of six hoops: ____________
- Bat: ____________
- Pack of 20 ropes: ____________
- Single rope: ____________
- Pack of five balls: ____________

Try to get close to the \$500 but do not overspend. When you have decided how to spend the money, write the list of the items and the prices neatly.

UNIT 4: TOPIC 1
Number patterns

Practice

1 Read the rules and complete each table.

a Start at 0. Increase by 1.5 each time.

Term	1	2	3	4	5	6	7	8	9	10
Number	0									

b Start at 5. Decrease by $\frac{1}{4}$ each time.

Term	1	2	3	4	5	6	7	8	9	10
Number	5									

2 Continue these patterns for the first 10 terms. Write a rule for each pattern.

a 30, 27.5, 25, 22.5, ______ Rule: ______

b 0, $\frac{1}{3}$, $\frac{2}{3}$, 1, ______ Rule: ______

c 10, 9.6, 9.2, 8.8, ______ Rule: ______

3 Look at each pattern and complete the table.

Pattern of sticks	Rule for making the pattern	How many sticks are needed?					
a	Start with 5 sticks. Increase the number of sticks by ______ for each new pentagon.	Number of pentagons	1	2	3	4	5
		Number of sticks					
b	Start with:	Number of diamonds	1	2	3	4	5
		Number of sticks					
c	Start with:	Number of pentagons	1	2	3	4	5
		Number of sticks					

Challenge

This flow chart shows rules for decreasing a number down to 1 using a divisor of 3.

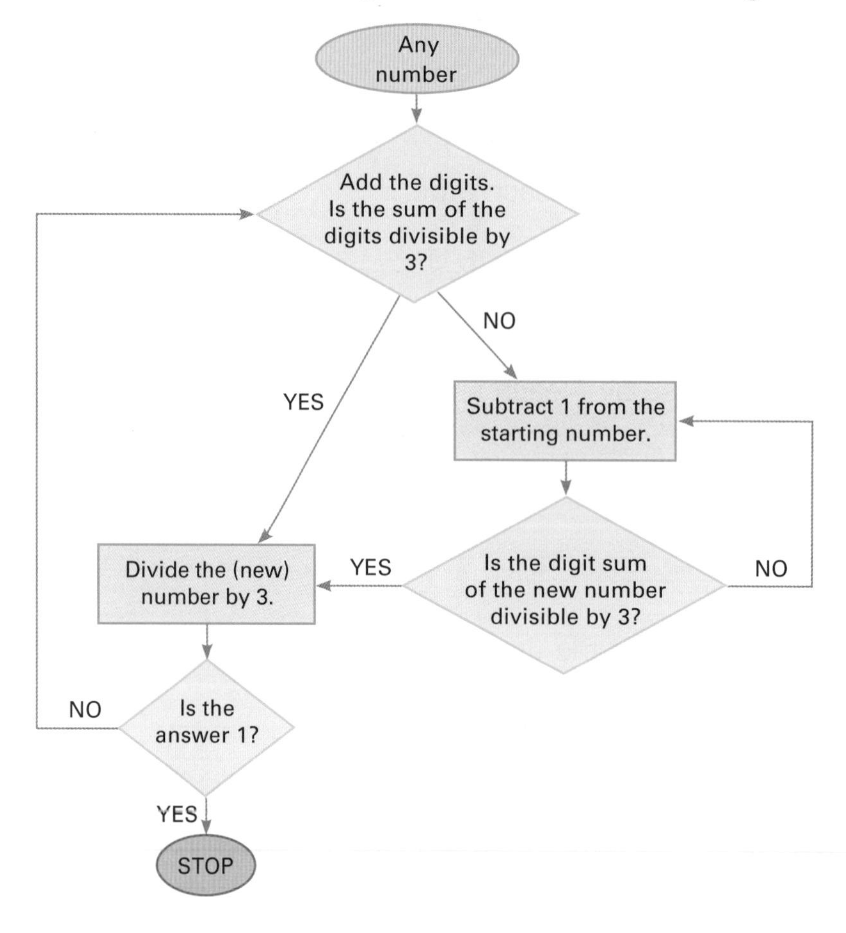

Following the rules, this is how 44 is reduced to 1:

- The digit sum of **44** is 4 + 4 = 8. 8 is not divisible by 3, so **44** – 1 = **43**.
- The digit sum of **43** is 4 + 3 = 7. 7 is not divisible by 3, so **43** – 1 = **42**.
- The digit sum of **42** is 4 + 2 = 6. 6 is divisible by 3, so **42** ÷ 3 = **14**.
- The digit sum of **14** is 1 + 4 = 5. 5 is not divisible by 3, so **14** – 1 = **13**.
- The digit sum of **13** is 1 + 3 = 4. 4 is not divisible by 3, so **13** – 1 = **12**.
- The digit sum of **12** is 1 + 2 = 3. 3 is divisible by 3, so **12** ÷ 3 = **4**.
- The digit sum of **4** is 4. 4 is not divisible by 3, so **4** – 1 = **3**.
- The digit sum of **3** is 3. **3** ÷ 3 = **1**
  STOP.

1 Follow the rules to take the following numbers to 1. Use a separate piece of paper.

a 35 **b** 362

2 Try using the rules with other, larger numbers. Use a separate piece of paper.

Mastery

1. On the previous page you followed rules to decrease a number to 1 using a divisor of 3. Think of rules for decreasing any number down to 1 using a divisor of 4. Make a flow chart to show the rules. Test your rules using 2-digit, 3-digit and 4-digit numbers. Use the flow chart below if it suits your rules. You may wish to make a draft version on spare paper first.

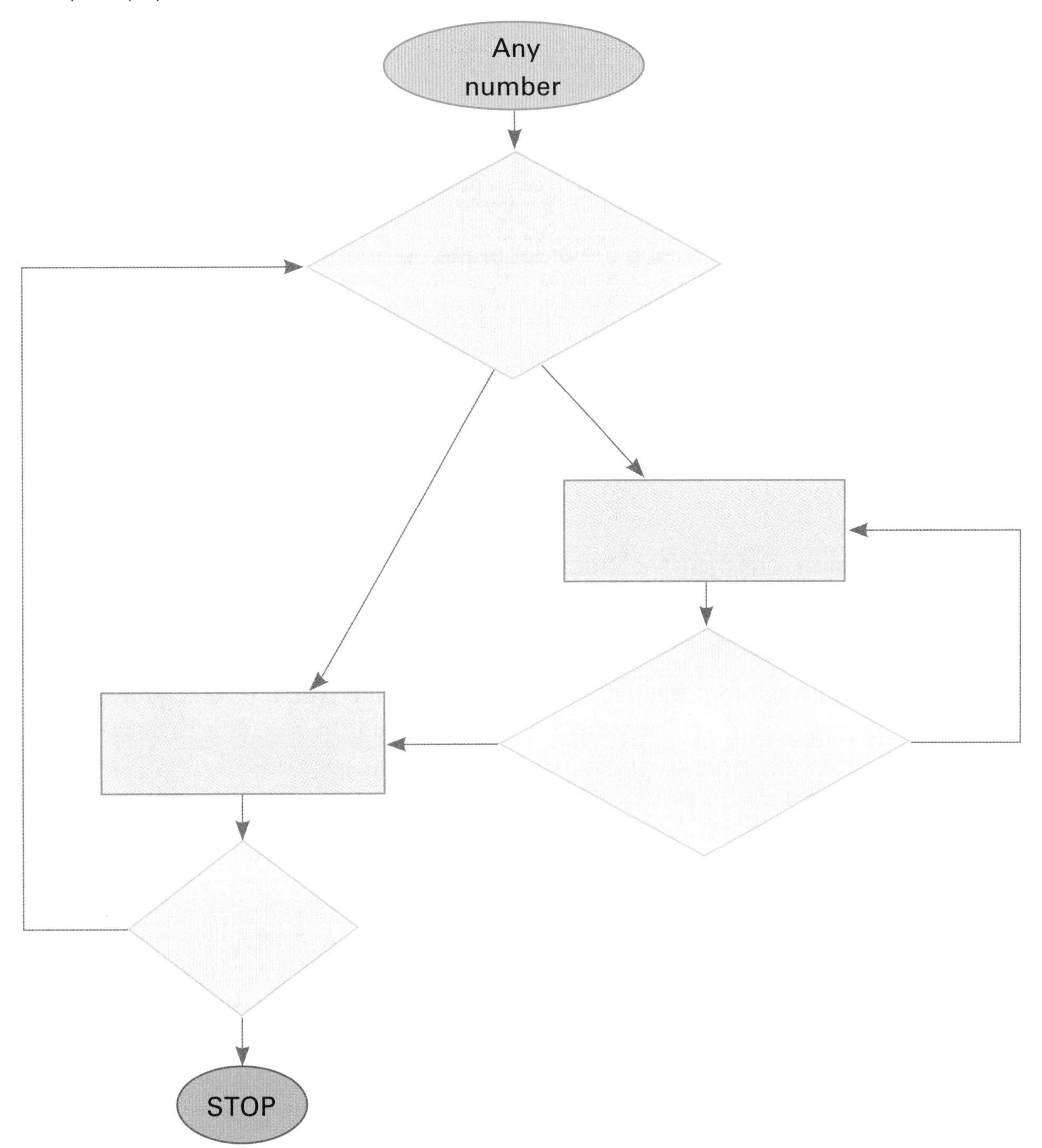

Number operations and properties

Practice

1 Decide how changing the order of the numbers affects the following problems. (Circle "does" or "does not".)

a $100 + 50 = ?$ $50 + 100 = ?$

Addition: Changing the order of the numbers **does/does not** give the same answer.

b $25 - 20 = ?$ $20 - 25 = ?$

Subtraction: Changing the order of the numbers **does/does not** give the same answer.

c $30 \times 2 = ?$ $2 \times 30 = ?$

Multiplication: Changing the order of the numbers **does/does not** give the same answer.

d $35 \div 7 = ?$ $7 \div 35 = ?$

Division: Changing the order of the numbers **does/does not** give the same answer.

---

2 Put these into an order that will help you to solve the problems easily.

a $25 + 18 + 15 =$ ____________________

b $20 \times 16 \times 5 =$ ____________________

c $47 + 37 + 13 =$ ____________________

d $500 \times 17 \times 2 =$ ____________________

---

3 There are three numbers in each of these subtraction and division problems. Find the answers, then change the order of the two numbers being subtracted or divided and solve the problems again. Circle whether or not changing the order gives the same answer.

a $43 - 16 - 23 =$ ____________________

b $81 \div 3 \div 9 =$ ____________________

c $69 - 25 - 19 =$ ____________________

d $96 \div 3 \div 8 =$ ____________________

Changing the order of the two numbers being subtracted or divided **does/does not** give the same answer.

Challenge

1 Fill in the gaps to make the equations balance.

a $24 \times 4 = \square + 35$

b $\square \div 5 = 75 \div 3$

c $48 \div 3 = 8 \times \square$

d $\square - 30 = 15 \times 4$

e $100 \div 5 = 19 + \square$

f $65 + 35 = \square \div 2$

g $25 \times 6 = \square \times 10$

h $\square - 20 = 15 \times 6$

i $120 \div 3 = 80 \div \square$

2 Find three **different** ways to balance each equation.

a

$5 \times 12 \times 2 =$

$5 \times 12 \times 2 =$

$5 \times 12 \times 2 =$

b

$22 + 45 + 8 =$

$22 + 45 + 8 =$

$22 + 45 + 8 =$

c

$120 \div 8 \div 3 =$

$120 \div 8 \div 3 =$

$120 \div 8 \div 3 =$

d

$240 \div 2 - 20 =$

$240 \div 2 - 20 =$

$240 \div 2 - 20 =$

e

$25 \times (2 + 8) =$

$25 \times (2 + 8) =$

$25 \times (2 + 8) =$

Mastery

1 Imagine a younger student is struggling with this mathematics problem:

$$9 \times 5 \times 4 \times 2$$

a Rewrite the number sentence in a way that would make it easier to solve, then use a mental strategy to solve it.

____

b Write an explanation that would help a younger student understand why rewriting the number sentence makes it easier to solve the problem.

____

____

2 Each of the following equations is made from the numbers 24, 8, 2 and a mystery number that is replaced by a diamond shape: ◆. In this number puzzle you need to find the value of the ◆. The ◆ has the same value in each equation. Remember to use the order of operations where necessary. Rewrite each equation using the mystery number instead of the ◆.

a 24 ÷ ◆ = 24 ÷ 2 – 8

____

b 24 × 2 = 8 × ◆

____

c (24 – ◆) + (24 ÷ ◆) = 8 + 8 + 8 – 2

____

d (24 ÷ 8) × 2 – ◆ = 24 – 8 – 8 – 2^2 × 2

____

3 **a** Write one or two other equations that use 24, 8, 2 and the same mystery number. ______________________________

____

b Try to write a number puzzle using three other numbers and a different mystery number.

UNIT 5: TOPIC 1
Length and perimeter

Practice

1 Write the length of the red lines in centimetres and millimetres, and in centimetres with a decimal. Use a ruler for questions c–g.

a

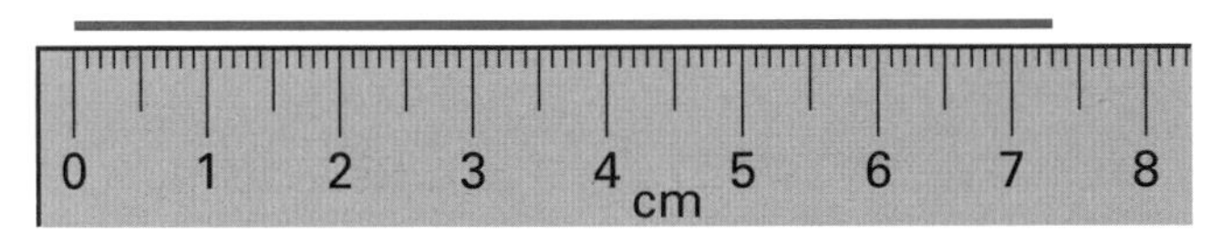

b

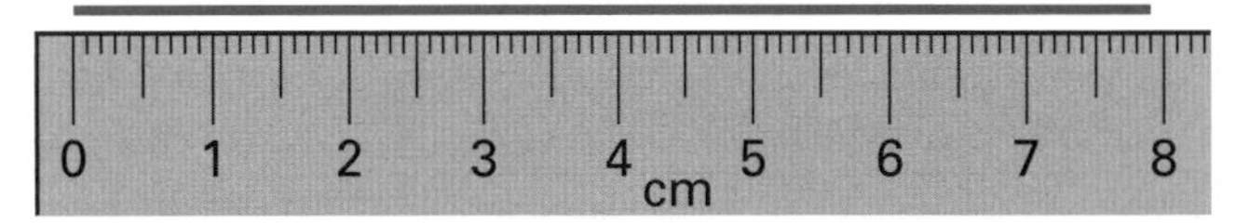

c

d

e

f

g

2 Calculate the perimeters of these rectangles without using a ruler.

a ________

8 cm

3 cm

b ________

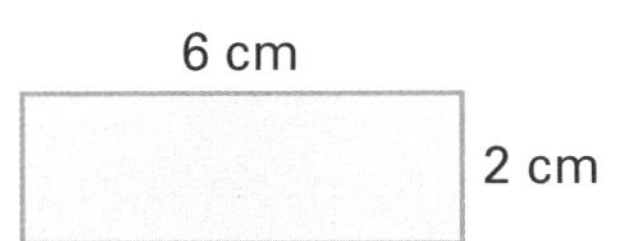

c ________

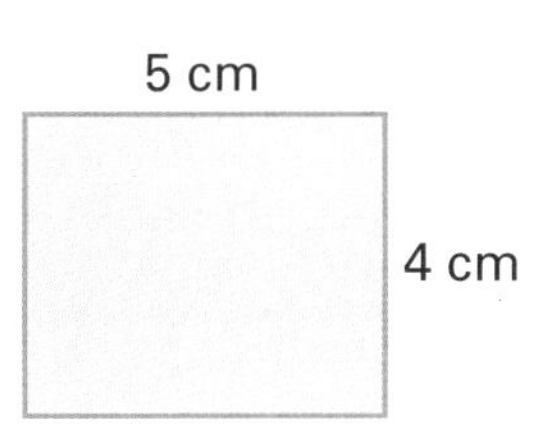

Complete these length-conversion tables.

3

	Centimetres	Millimetres
a	5 cm	
b		55 mm
c	27 cm	
d	3.8 cm	
e		22 mm

4

	Metres	Centimetres
a	3 m	
b		250 cm
c	7.35 m	
d		125 cm
e	$3\frac{1}{2}$ m	

5

	Kilometres	Metres
a	5 km	
b		3500 m
c		7250 m
d	4.75 km	
e		1100 m

Challenge

1 Choose an appropriate measuring tool to measure the following objects in your classroom. Use two different units of length to show the length of each object.

a An eraser: ________ ________

b The height of the classroom door: ________ ________

c The width of the whiteboard: ________ ________

d The width of the classroom: ________ ________

2 Measure to find the perimeters of these shapes. Write the answer in millimetres and in centimetres with a decimal.

a

Perimeter:

________ mm

________ cm

b

Perimeter:

________ mm

________ cm

c

Perimeter:

________ mm

________ cm

d

Perimeter:

________ mm

________ cm

3 Draw an equilateral triangle that has a perimeter of 135 mm.

4 Draw a line that is 210 mm long inside this oval.

Hint: A line does not need to go in just one direction!

Mastery

1 The wedge-tailed eagle is also known as the eaglehawk. It has the greatest wingspan of all eagles, measuring almost 3000 mm across. If you were researching the wingspans of eagles, perhaps you would not expect the unit of length to be millimetres. This table shows the eagles that have some of the biggest wingspans.

Type of eagle	Wingspan (mm)	Wingspan (cm and mm)	Wingspan (cm)	Wingspan (m)
Wedge-tailed eagle	2840 mm			
Golden eagle	2805 mm			
Martial eagle	2625 mm			
Sea eagle	2490 mm			
Bald eagle	2275 mm			

Complete the table by converting the lengths into the various units.

2 If you were using the information in the table to write a report about eagles, which unit of length would you use to describe the following? Give a reason for your answer.

a The wingspan:

b The length of a claw:

c The diameter of the eye:

3 With your teacher's permission, carry out some research on some record-breaking lengths. These could range from gigantic sea creatures, dinosaurs or aircraft, to the tiniest model trains. Decide on the unit of length to use and illustrate it if you wish.

Area

Practice

1 Find the area of these rectangles:

a

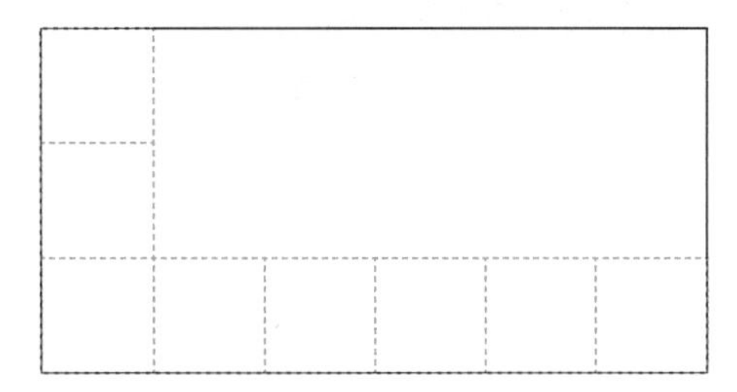

Area = ________

b

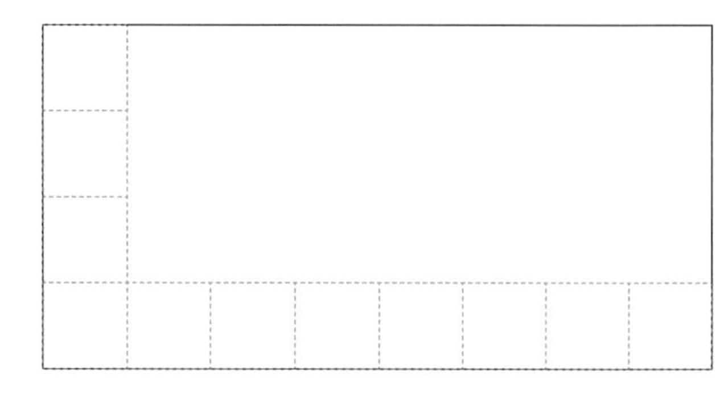

Area = ________

c

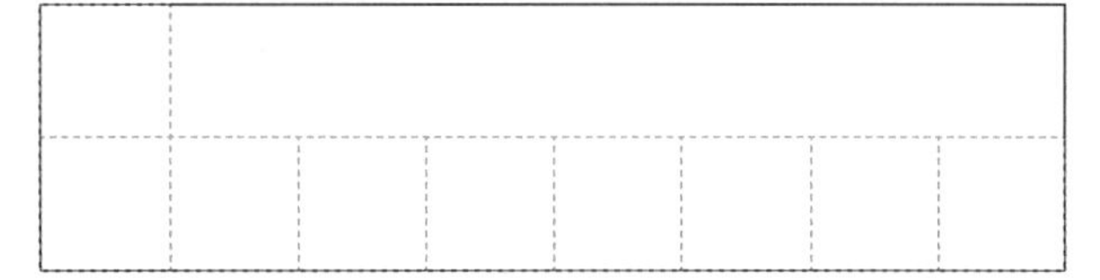

Area = ________

d

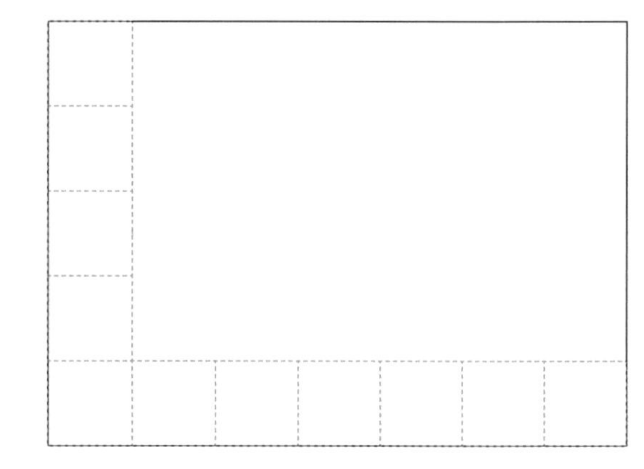

Area = ________

e

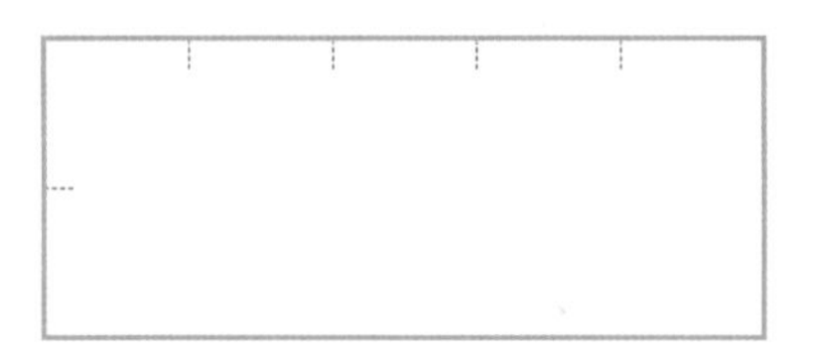

Area = ________

f

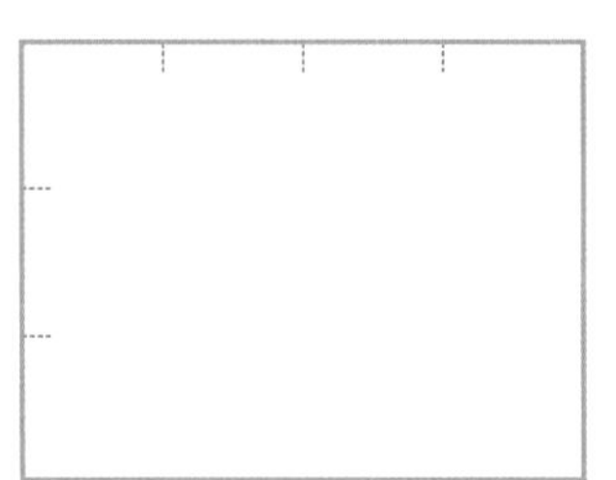

Area = ________

g

Area = ________

h

Area = ________

Challenge

Bathroom/laundry

Living room

Bedroom

1 This is the floor plan of a small apartment. It is drawn to scale so that 1 centimetre on the drawing is the same as 1 metre in real life. You will need to measure the diagram to answer the following questions.

What is the area of:

a the living room? ________

b the bedroom? ________

c the bathroom/laundry? ________

d the whole apartment? ________

---

2 Choose an appropriate measuring tool to find the measurements you need to calculate the area of each shape below.

a

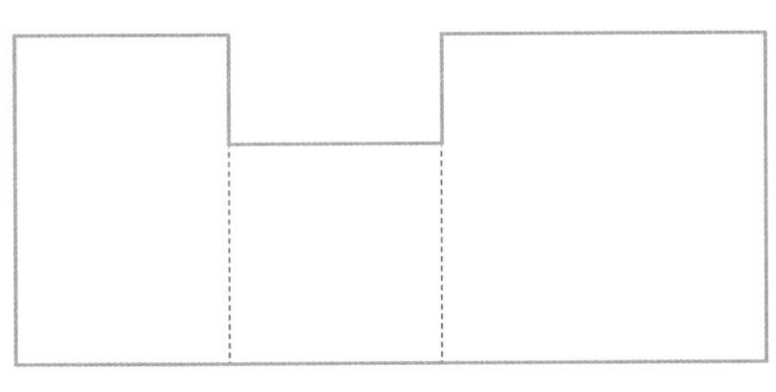

Area = ________

b

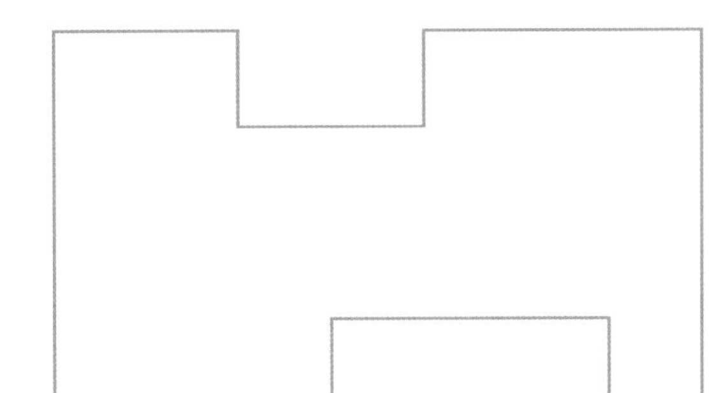

Area = ________

---

3 This is a diagram of a garage/storage area. It is drawn to scale so that 1 centimetre on the drawing is the same as 1 metre in real life.

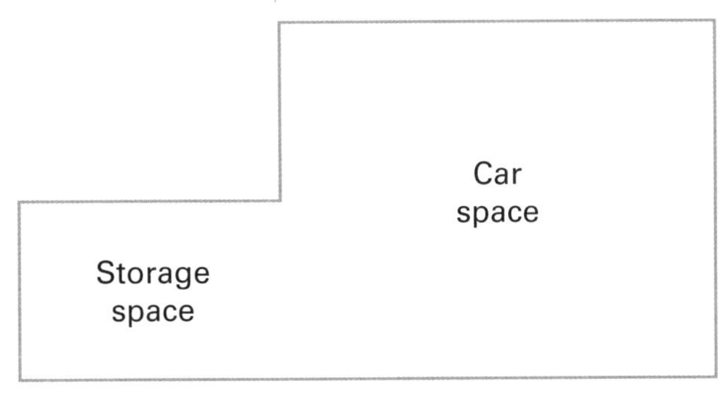

a What is the total floor area? ________

b What is the area of the car space? ________

c What is the area of the storage space? ________

Mastery

1 To measure a very large area we use square kilometres (km^2). This map shows the states and territories in Australia. The state of New South Wales (NSW) has an area of about 800 000 km^2. If this number were written in square metres it would be hard to read: 800 000 000 000 m^2. (There are a million square metres in a square kilometre.)

Legend:

WA	Western Australia
SA	South Australia
TAS	Tasmania
VIC	Victoria
ACT	Australian Capital Territory
NSW	New South Wales
QLD	Queensland
NT	Northern Territory

a Match the other areas in the table to the correct states and territories. You will probably need to do some research to complete the table. (Note: all the areas are rounded, so your research may result in an area that does not exactly match the rounded number in the table.)

State/territory	Area (rounded)
	2.5 million km^2
	1.7 million km^2
	1.3 million km^2
	980 000 km^2
NSW	800 000 km^2
	225 000 km^2
	68 000 km^2
	2000 km^2

b Add the rounded totals of each state and territory to find the approximate area of Australia.

____

2 You saw on the map of Australia that the states are not rectangles. How do you think people worked out the areas? Find a way to work out the area of something that has angles other than right angles. You could begin by trying to find the area of this triangle.

3 In the diagrams on page 57, all the rooms were an exact number of metres. It is often not like that in real life. Find the area of a real room, such as your classroom or a room in your house. Begin by finding the approximate area. (Round the length and width to the nearest metre.) When you have done that, try to find a way to work out the exact area.

UNIT 5: TOPIC 3
Volume and capacity

Practice

Capacity means the amount that something can contain.

1 Circle the most likely capacity of the following.

a

120 mL   12 mL
120 L   12 L

b

2 L   20 mL
200 mL   2 mL

c

45 mL   45 L
450 L   450 mL

d

5 mL   5 L
50 mL   50 L

2 Write the volume of each centimetre cube model.

a

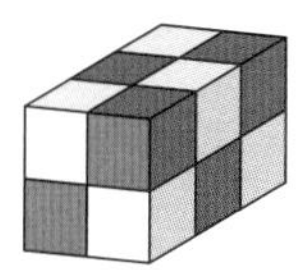

Volume = ________ cm^3

b

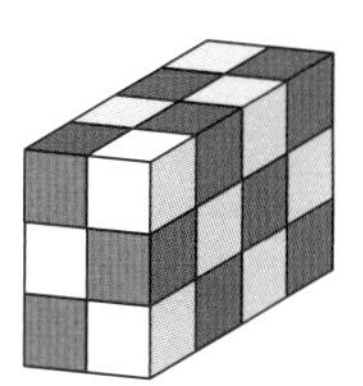

Volume = ________

c

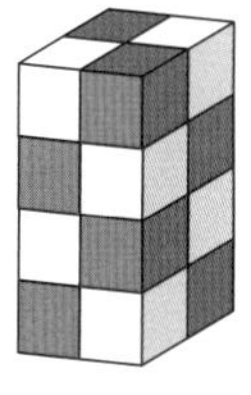

Volume = ________

3 a How could you explain to a younger person that the volume of this model is 18 cm^3?

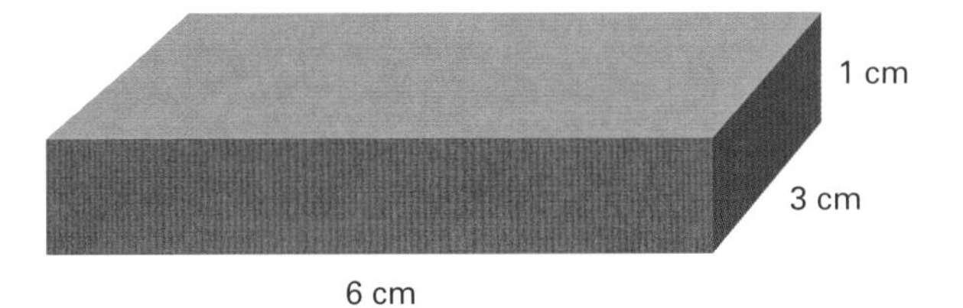

____

b What would the volume be if the height were 4 cm? ________

4 a What is the volume of the transparent box? ________

b What would the volume be if the width were 3 cm? ________

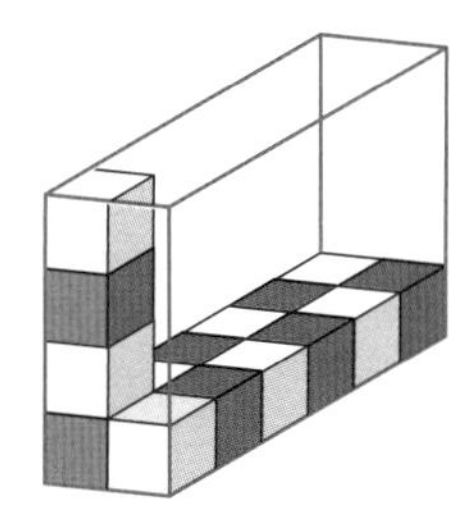

Challenge

1 **a** Write the capacity of this bottle in as many different ways as you can.

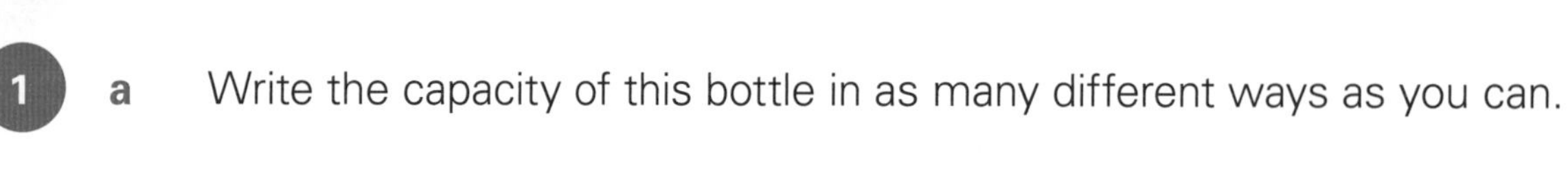

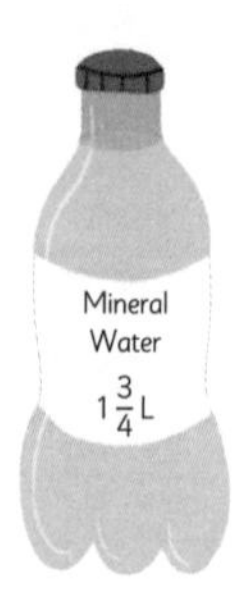

b How many millilitres would three bottles hold? ________

---

2 Write the volume of each rectangular prism.

a

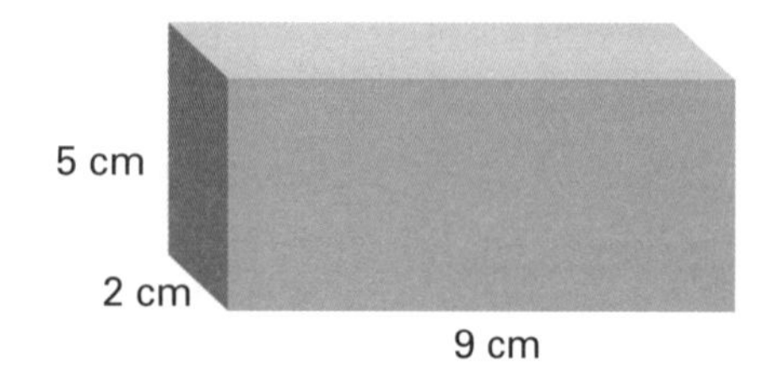

Volume = ________

b

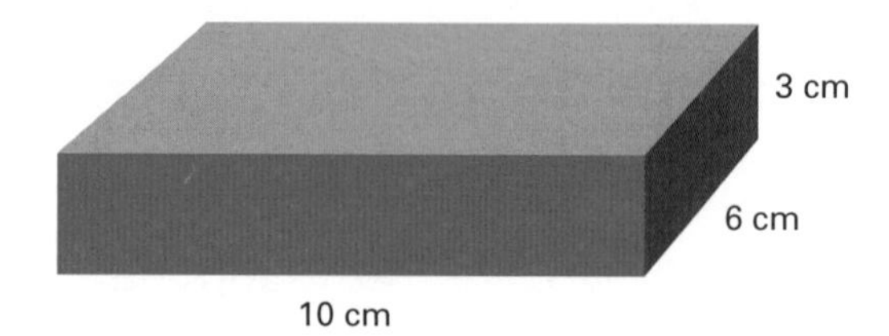

Volume = ________

c

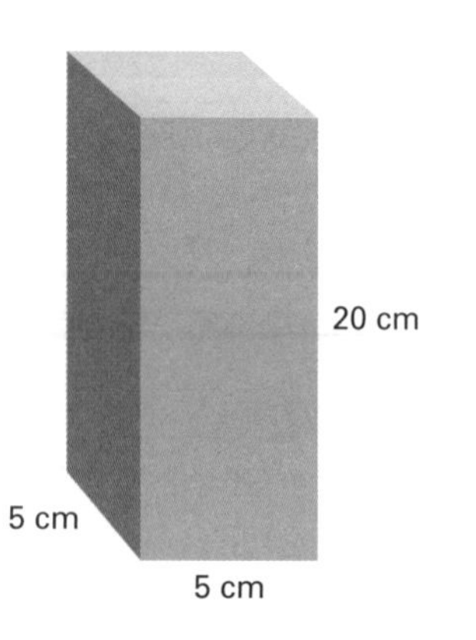

Volume = ________

d

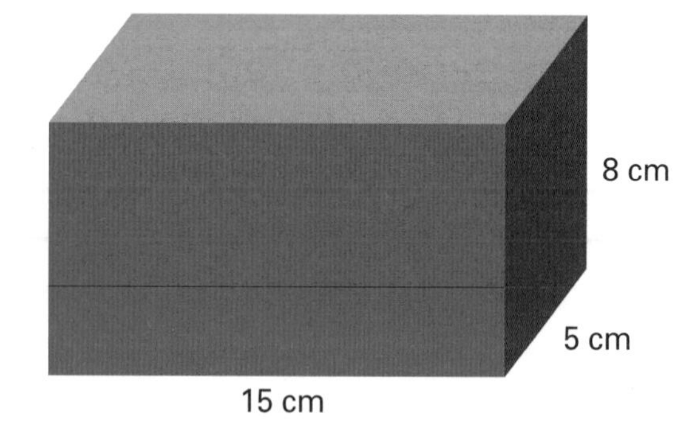

Volume = ________

e

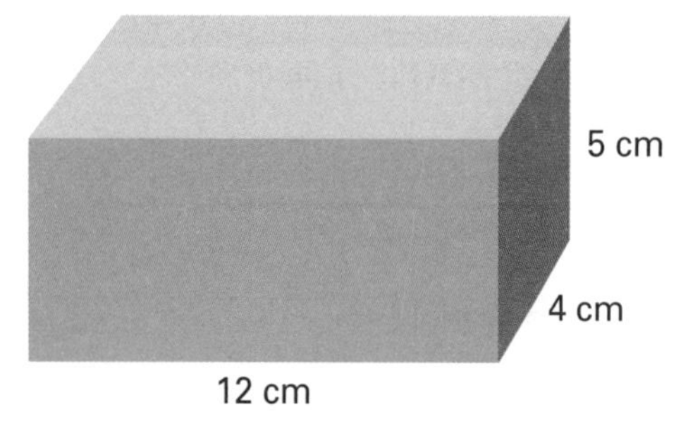

Volume = ________

f

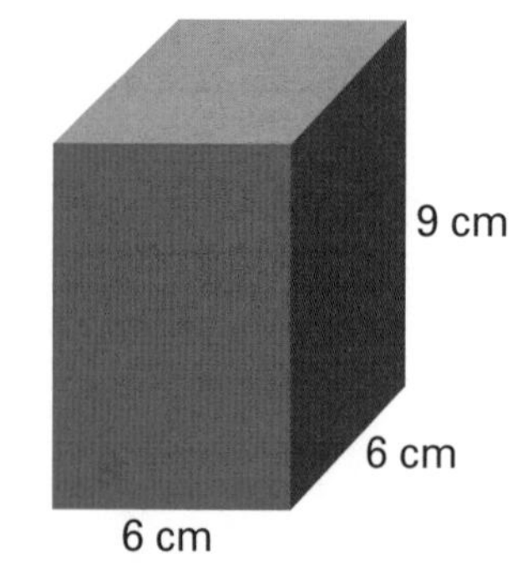

Volume = ________

---

3 How much water would spill out of a full bucket if the following objects were placed in it?

a A centimetre cube measuring 5 cm by 10 cm by 5 cm: ________

b A 10 cm × 10 cm × 10 cm cube of concrete: ________

c A paving brick 250 mm long by 150 mm wide by 100 mm deep: ________

Mastery

1. People who live in countries with low rainfall need to be "water-wise" because there are times of drought. One way to be water-wise is to make sure that we do not waste water. Did you know, for example, that if someone leaves the tap running while they brush their teeth, around 50 litres of water run down the drain?

Carry out some research to collect some facts about water usage. Think of a few **Dos** and **Don'ts** in connection with the way people use water. You could choose one or two to turn into a poster to encourage people in your school to be water-wise. List some of your hints here.

2. Perhaps your research included information about water-wastage from a dripping tap.

How much water would a dripping tap waste? The answer is probably not much while you stood watching it. However, imagine if it dripped all day and all night for days and weeks! Find out how much water would be wasted.

Collect water from a dripping tap for five minutes. If you don't have a dripping tap, you could set one to drip. (Don't worry, you can be water-wise and use the water afterwards.)

Choose an appropriate measuring tool to measure the amount of water you collected and use a calculator to find out how much water a dripping tap would waste in an hour, a day, a week or even a year.

Write a short report about your findings.

Mass

Practice

1 Convert between these units of mass.

	Tonnes	Kilograms
a		3000 kg
b	5 t	
c	2.5 t	
d		4250 kg
e	2.75 t	

	Kilograms	Grams
a	7 kg	
b		4000 g
c		1250 g
d	5.75 kg	
e	0.25 kg	

	Grams	Milligrams
a	1 g	
b		2000 mg
c		7500 mg
d	1.25 g	
e	0.75 g	

2 What is something that is likely to have its mass shown in:

a milligrams? ____________ b grams? ____________

c kilograms? ____________ d tonnes? ____________

3 Write the mass of each box in grams.

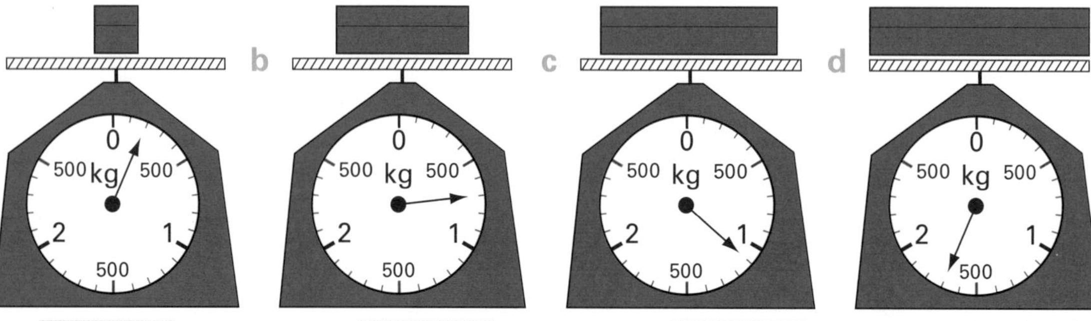
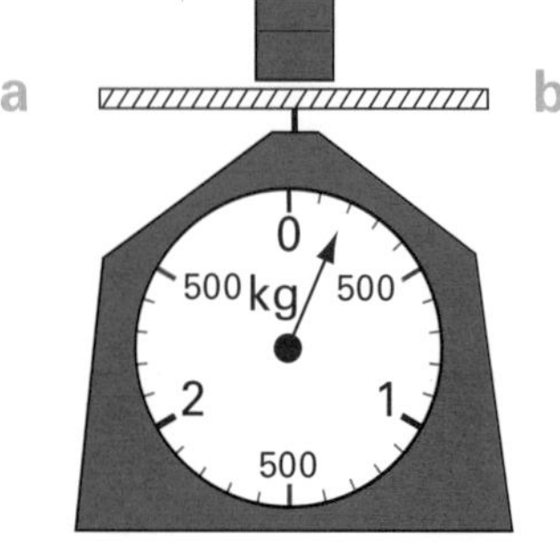

Mass: ____

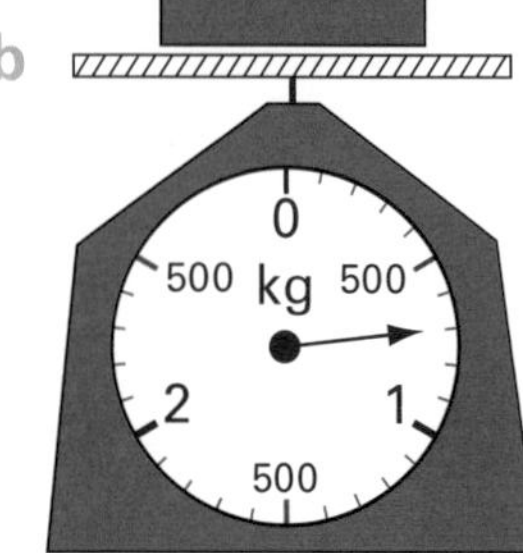

Mass: ____

c

Mass: ____

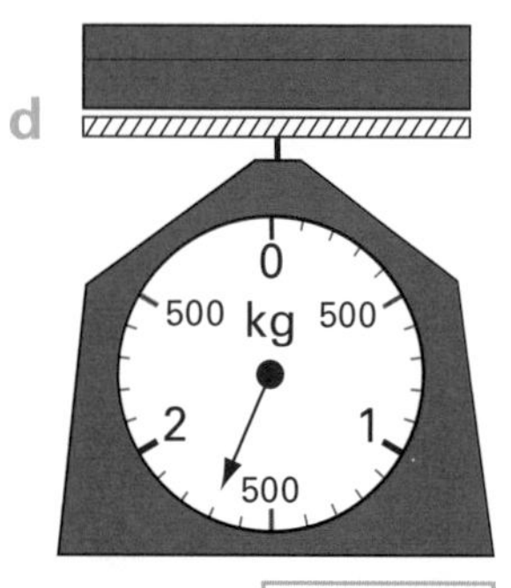

Mass: ____

4 Look carefully at these scales and write the masses in kilograms and grams and in kilograms with a decimal.

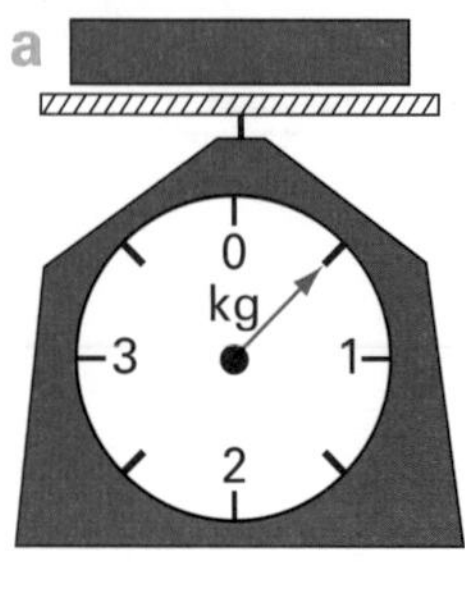

____ kg ____ g

____ kg

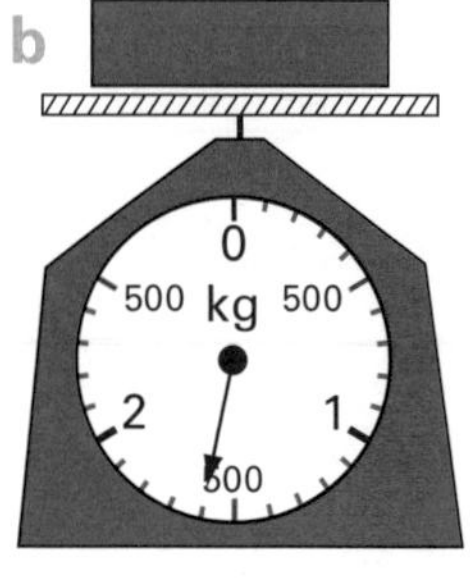

____ kg ____ g

____ kg

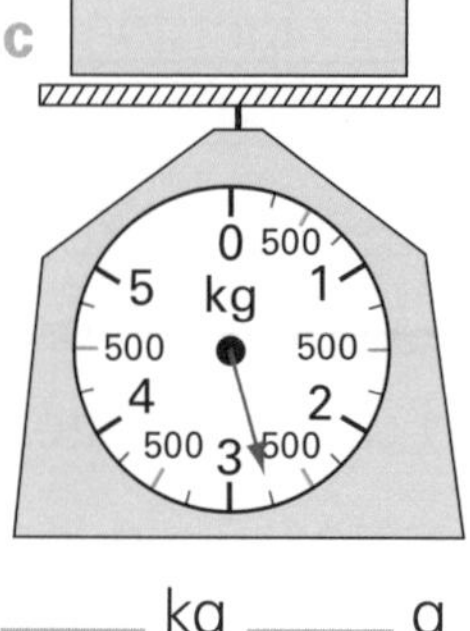

____ kg ____ g

____ kg

d

____ kg ____ g

____ kg

Challenge

1 Draw pointers on the scales to show the mass of each box.

a 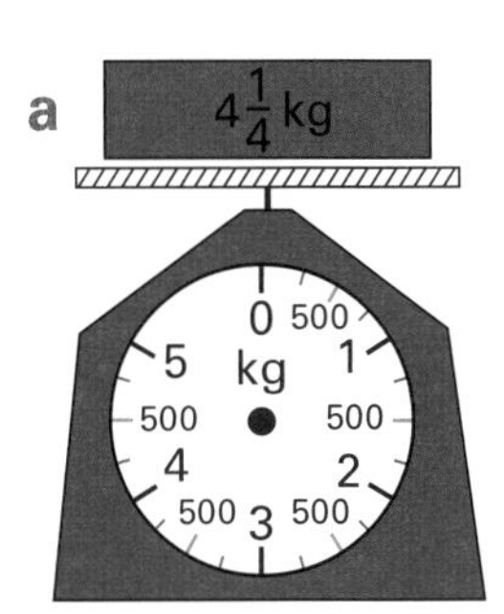

b 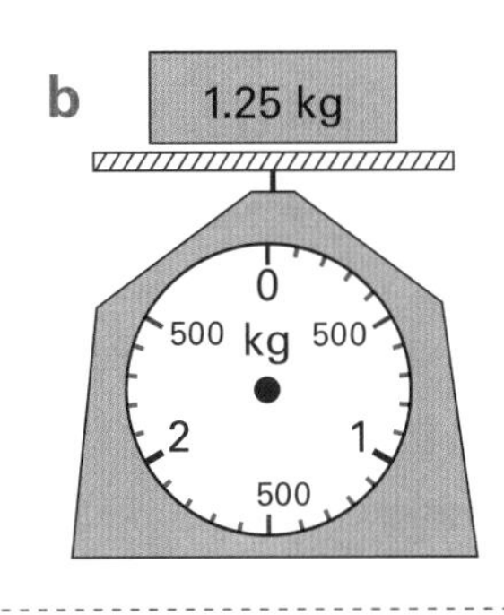

c 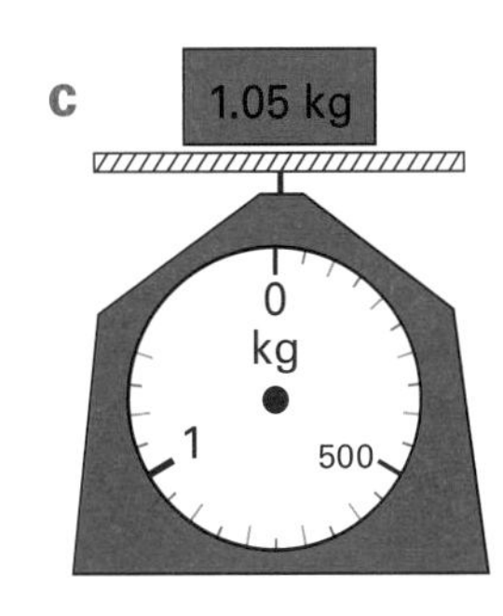

d 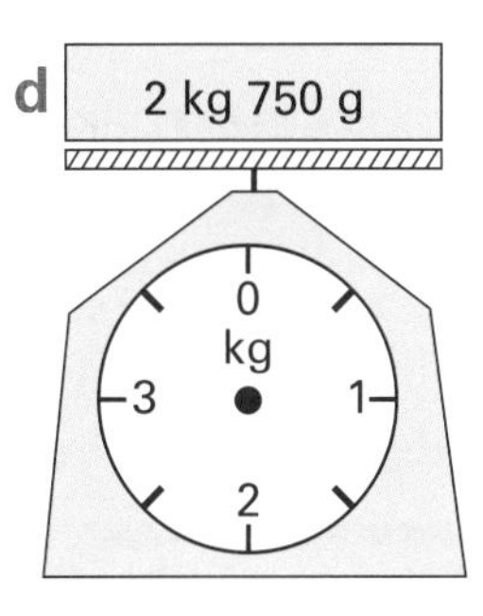

2 Write the masses of the boxes in question 1 in as many different ways as you can.

Box A ______________________ Box B ______________________

Box C ______________________ Box D ______________________

3 These children are on a junior basketball team.

Eva
29.1 kg

Alex
29.35 kg

Mo
$28\frac{3}{4}$ kg

Dee
30 kg 200 g

Billy
28.1 kg

Jo
$28\frac{1}{2}$ kg

a Write the names in order from lightest to heaviest.

____

b Last year Alex was $1\frac{1}{2}$ kg lighter. What was his mass then? ______________

c What is the mean (average) mass of a player on the team? ______________

d Which player is closest to the average mass? ______________

e How many grams lighter than Dee is Eva? ______________

f Whose combined mass is closest to 60 kg? ______________

4 Audrey is going on a plane trip. She has a baggage allowance of 20 kg, and already has 18 kg in her bag. She wants to take some drinks and snacks. When she puts them on a scale, the combined mass is exactly 2 kg. Write a possible mass for each item. Make sure the total is 2 kg.

Item	Mass	Item	Mass
Small bottle of water		Carrot sticks	
Large bottle of water		Apple	
Bottle of juice		Bunch of grapes	

Mastery

1 The blue whale is the heaviest sea creature in the world. If someone told you that the average mass of an adult blue whale was 600 grams less than 190 tonnes, what would that make the mass of the blue whale in kilograms? ______

2 All the heaviest sea creatures are whales. This table shows some of the heaviest whales, apart from the blue whale. It is in alphabetical order.

Type of whale	Average mass
Bowhead whale	99.79 tonnes
Grey whale	39 999 kg 600 g
Humpback whale	39.938 tonnes
North Atlantic right whale	95 508 kg
North Pacific right whale	89.412 tonnes
Southern right whale	86 tonnes 364 kg

a Write the list of six whales in order from heaviest to lightest.

______

b What is the difference between the heaviest and lightest right whales? Write your answer in three ways: in kilograms, in tonnes and kilograms, and in tonnes with a decimal.

______

c If a blue whale were placed on one side of a giant set of scales, which other two whales would be closest to balancing it?

______

d How much heavier than the two whales from the answer to question 2c would the blue whale be? ______

3 Make your own list of the heaviest/lightest of a subject that interests you. It could be anything, from fish or flightless birds to trucks or spacecraft. Research your chosen topic. On a separate piece of paper, make a table placing the masses in order. Then make up some questions for others to answer, making sure that you know the answers yourself. Illustrate your work where possible.

Time

Practice

1 Use am/pm and 24-hour times to complete this timeline.

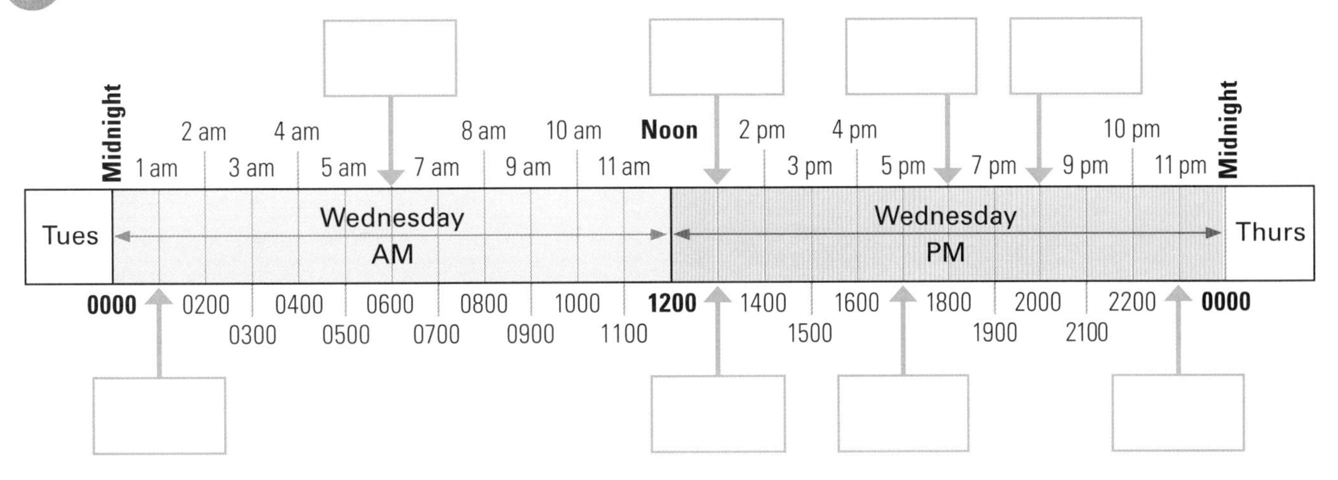

2 Convert between these am/pm and 24-hour times.

a	11 pm	b	1130	c	6:35 pm	d	0825
	______		______		______		______
e	2140	f	7:22 pm	g	4:44 am	h	0005
	______		______		______		______

3 Write these events as 24-hour and am/pm times.

	Event	am/pm time	24-hour time
a	The time I get home on a school day		
b	The time I eat breakfast at the weekend		
c	The time I wake up on a school day		
d	The time I eat lunch on a school day		

4 Fill in the gaps to show these times four different ways. Remember to look for the red pm indicator.

a

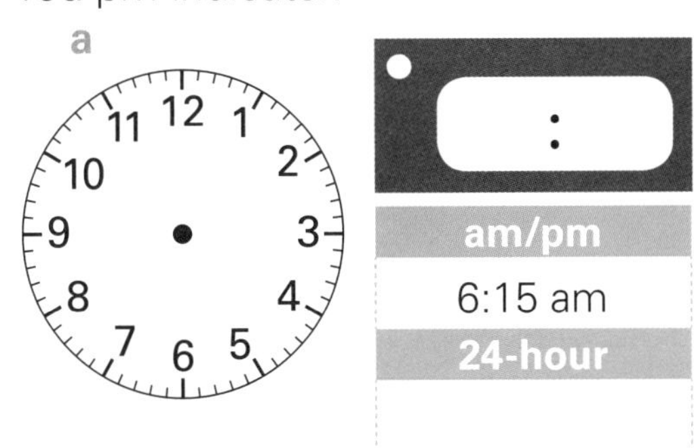

b

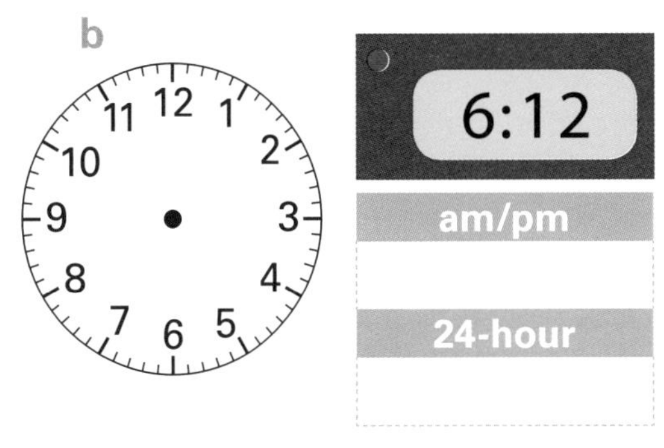

Challenge

1 This is the timetable for one day at a school camp.

a Fill in the 24-hour times in the answer spaces on the chart.

b How long do the students spend tidying their rooms? ______________

c At what time is dinner? (Use 24-hour time.) ______________

d How long does dinner time last?

____

e When does quiet time begin and end? (Use 24-hour time.)

____

f At what time is lunch? (Use 24-hour time.) ______________

g When does the tennis session end? (Use am/pm time.) ______________

h When lunchtime finishes, how long do the students have to wait for dinner?

____

24-hour time	Tuesday	am/pm time
[ ]		7:45 am
	Breakfast	
[ ]		8:15 am
	Tidy rooms	
		8:35 am
	Roll call	
[ ]		8:45 am
	Archery (90 minutes)	
[ ]		10:25 am
	Snack time	
		10:45 am
	Tennis (70 minutes)	
[ ]		noon
	Lunch	
[ ]		12:45 pm
	Free time	
		1 pm
	Kayaking (80 minutes)	
[ ]		2:30 pm
	Quiet time	
		2:40 pm
	Team games then free time	
[ ]		6:15 pm
	Dinner	
[ ]		7:05 pm
	Quiz time	
[ ]		8:15 pm
	Bed time	
[ ]		8:50 pm
	Lights out	

i Looking at the timetable, suggest a suitable time for an afternoon snack. (Use 24-hour time.) ______________

j How long is it from the start of team games to dinner time? ______________

k If you were in charge of the timetable, how would you divide up the time that is shown as "Team games then free time"?

Mastery

1

Town	Time of arrival/ departure
Tinytown	2200
Smallville	
Bigtown	
Medium City	0400
Largetown	
Great City	0540
Hugeville	

There are some gaps in this timetable for the bus from Tinytown to Hugeville. Here is the missing information:

- It takes three hours and 10 minutes to get from Tinytown to Smallville.
- Bigtown is two hours and 20 minutes before Medium City.
- Largetown is an hour and a quarter after Medium City.
- The last part of the journey takes 30 minutes.

a Use the information to complete the timetable. Use 24-hour times.

b Which two towns is the bus between at midnight? ____________________

c Which is the shortest part of the journey? ____________________

d How long does it take for the whole journey? ____________________

---

2 Times around the world are calculated in relation to what the time is in Greenwich, London. The time in Greenwich is known as Greenwich Mean Time or GMT. For example, Sydney is in the GMT +10 time zone, which means it is 10 hours ahead of Greenwich. When it's 1 am in Greenwich, it's 11 am in Sydney.

Calculate what the time is in each of the following cities if GMT is 3 pm.

a New York: GMT –5 __________

b Hong Kong: GMT +8 __________

c Auckland: GMT +12 __________

d Honolulu: GMT –10 __________

e Kabul: GMT +4:30 __________

f Istanbul: GMT +3 __________

---

3 Conduct some research to find the time in five other cities around the world when it is 3 pm GMT.

2D shapes

Practice

1 Look at each 2D drawing below:

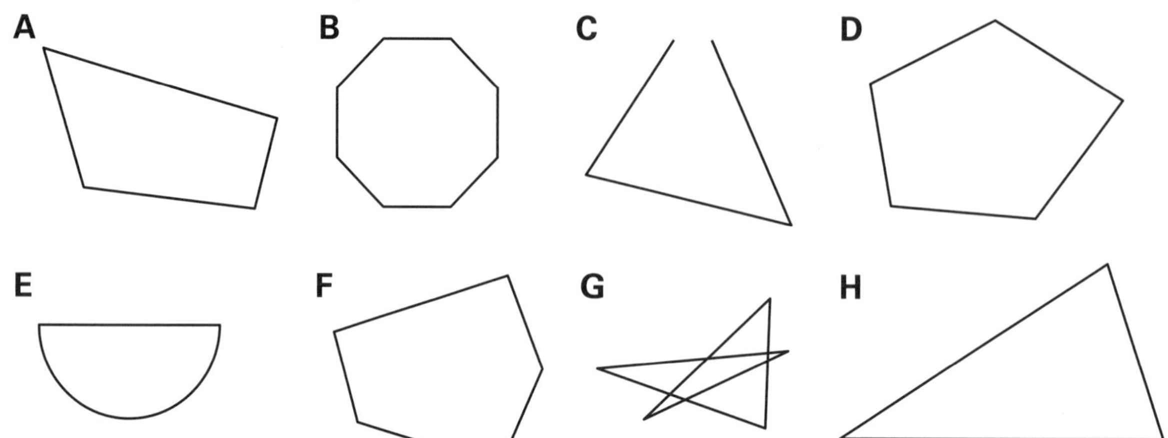

a If it **is** a polygon, write its name inside the shape.

b If it **is not** a polygon, explain why.

Drawing ______ is not a polygon because ______________________

Drawing ______ is not a polygon because ______________________

Drawing ______ is not a polygon because ______________________

2 Shade the polygons that are **regular**:

a

b

c

d

e

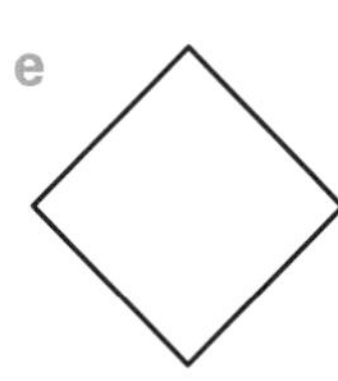

f

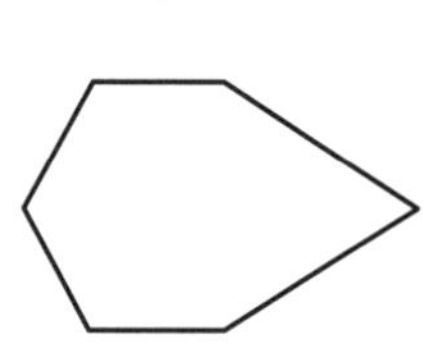

g

h

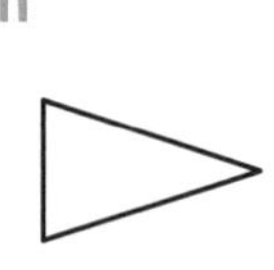

i

j

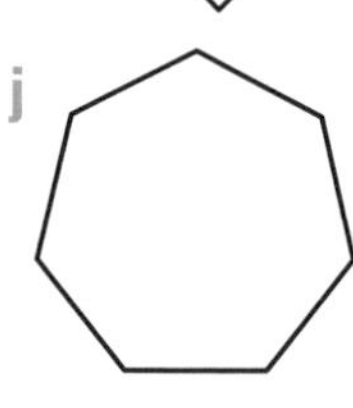

3 Decide whether the triangles below are **scalene**, **equilateral** or **right-angled**. Write your answers in the spaces provided.

a

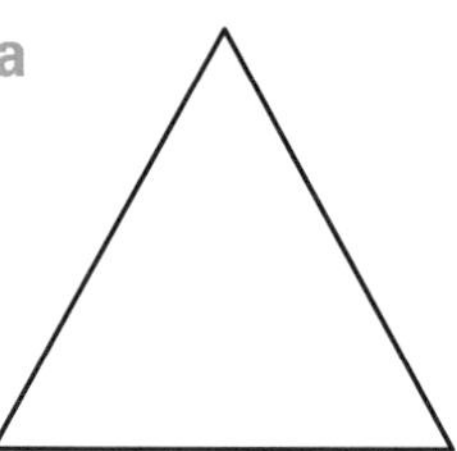

b

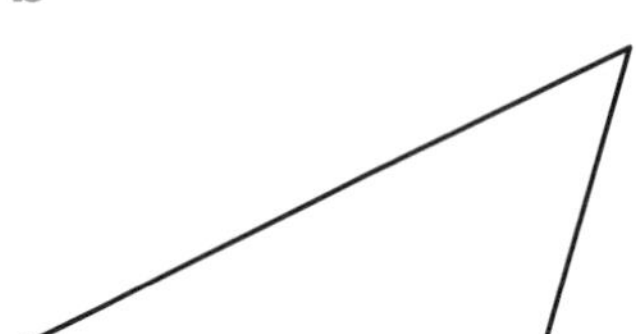

c

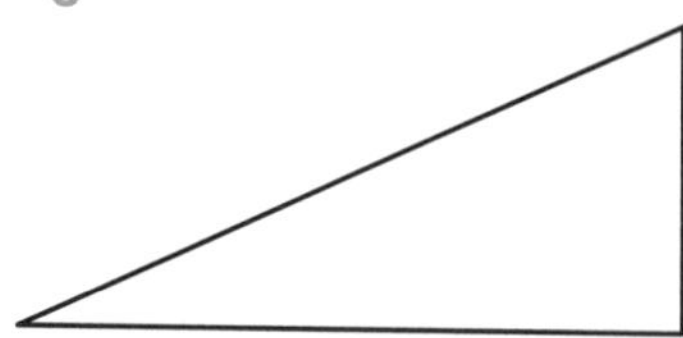

______________ ______________ ______________

Challenge

1 Draw the following quadrilaterals. Be as accurate as possible.

a Draw a rectangle. **b** Draw a parallelogram. **c** Draw an irregular quadrilateral.

2 How do you know that:

a this is a square? **b** this is a trapezium? **c** this is a rhombus?

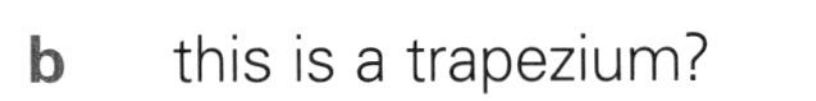

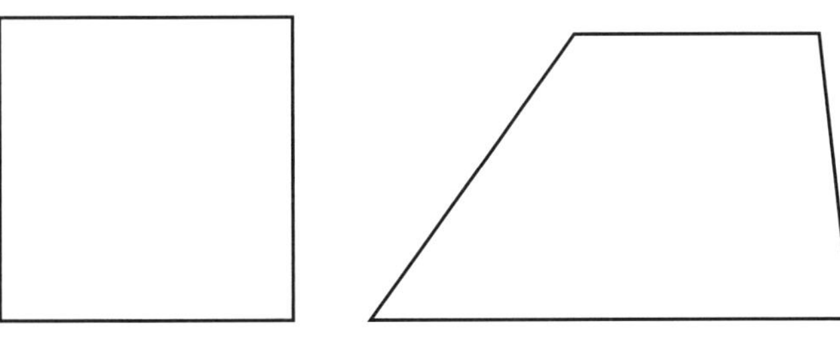

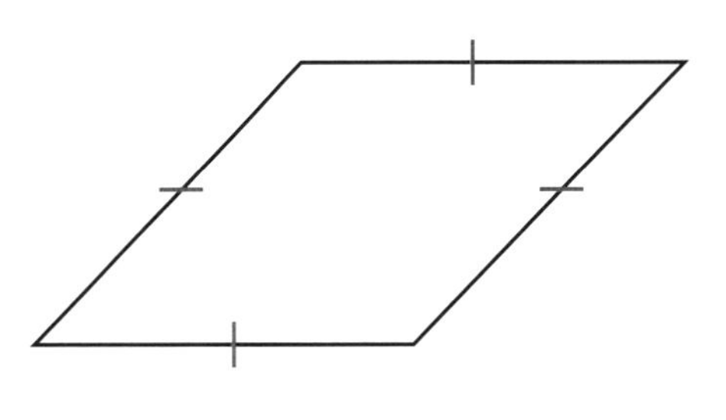

3 Identify the polygons from their descriptions.

a This polygon has three sides, one right angle and no sides of equal length. It is: ______

b This polygon is a parallelogram. It has four right angles. It has two long sides of equal length and two short sides of equal length. It is: ______

c This polygon has four sides. It has no right angles. The opposite sides are parallel and of equal length. It is: ______

d This polygon has five unequal angles. It is: ______

4 Write some information about this 2D shape.

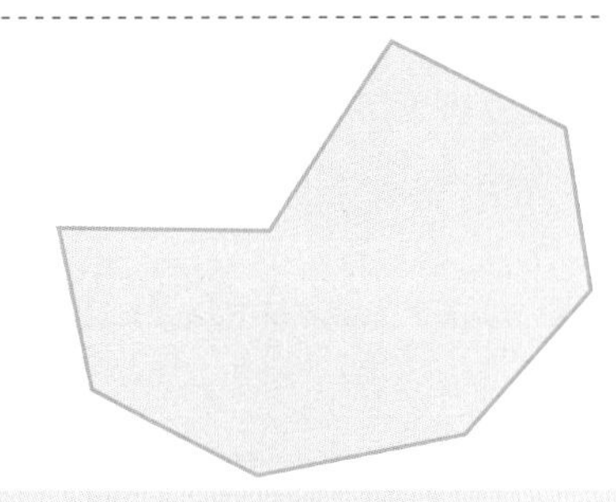

Mastery

1. This is a tangram puzzle.

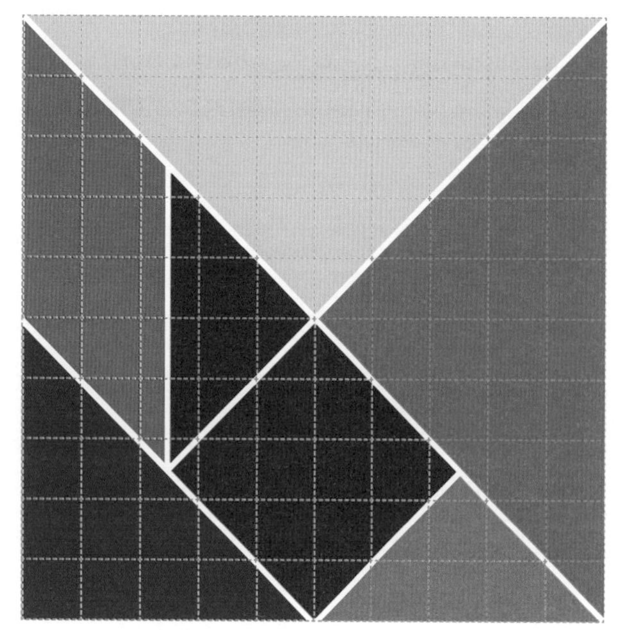

List the shapes that make up the coloured puzzle.

____

____

2. Make your own tangram puzzle by copying the pattern onto some centimetre square grid paper.

- Colour the shapes as you wish.
- Glue the puzzle onto card (optional).
- Cut out the seven shapes.

3. For centuries, people have used the shapes from tangram puzzles to make pictures, ranging from people to animals to buildings. Use the shapes to design some pictures. You could begin by finding out how the shapes fit together to make these two pictures. Draw around the shapes to make pictures and/or patterns.

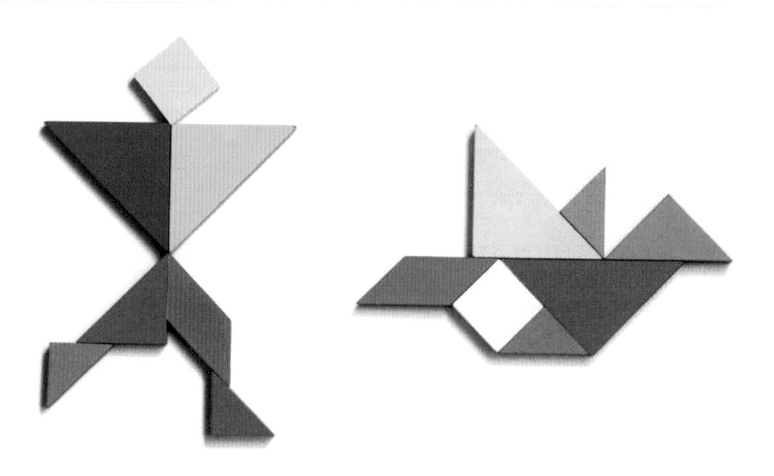

4. Nick made a picture using at least three different shapes with a total of exactly 14 angles. Draw what the picture might have looked like and list the shapes he used.

UNIT 6: TOPIC 2
3D shapes

Practice

1 Look at each of the following 3D shapes.

a Write the name under each 3D shape.

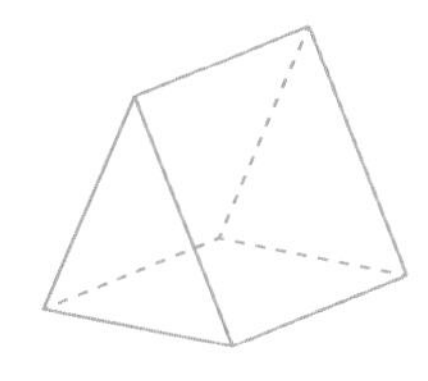

A ____________

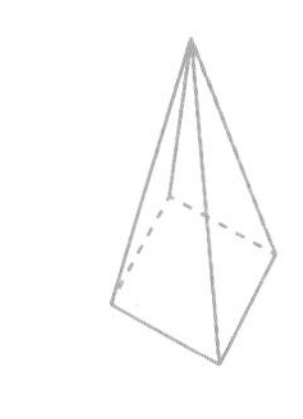

B ____________

C ____________

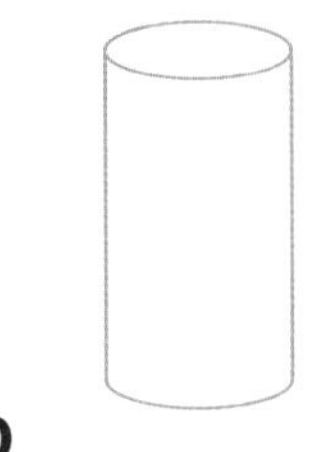

D ____________

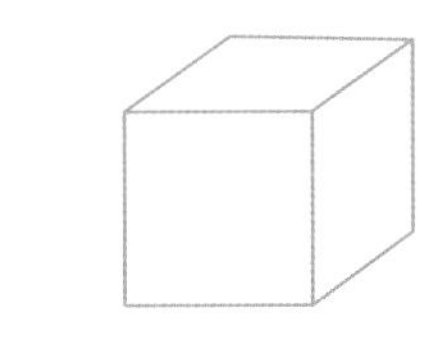

E ____________

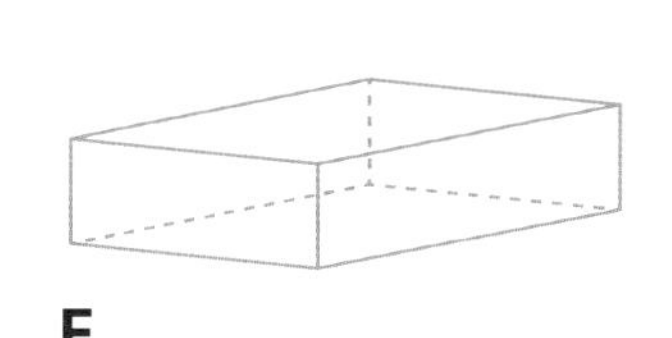

F ____________

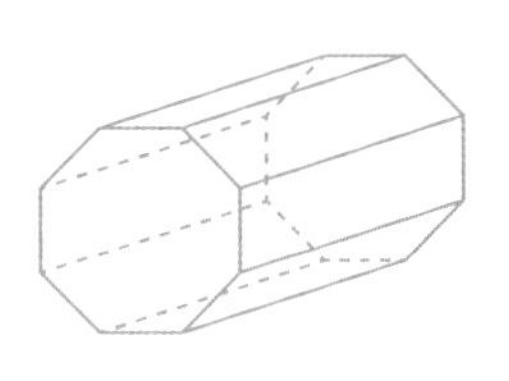

G ____________

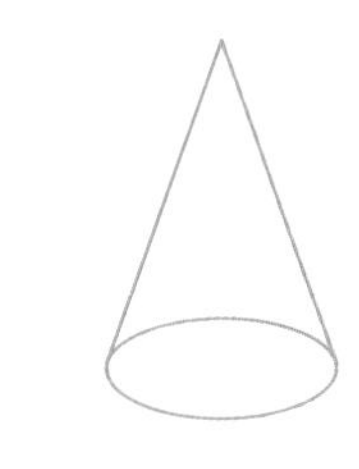

H ____________

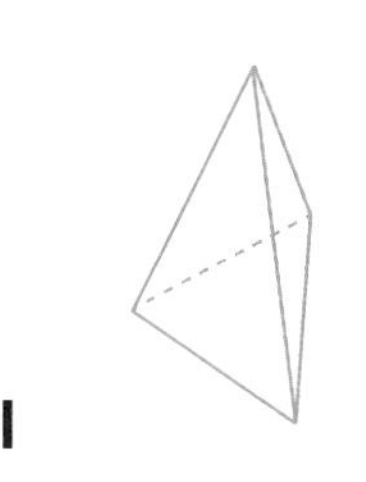

I ____________

b Some of the 3D shapes are **not** polyhedra. Which ones? Explain why.

Objects ____________ are not polyhedra because ____________

____

2 One of these objects is a prism and one is a pyramid. Write the name of each and explain how you know whether it is a prism or a pyramid.

A

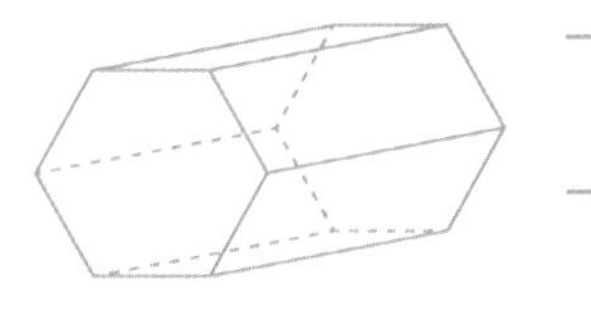

____

____

____

____

B

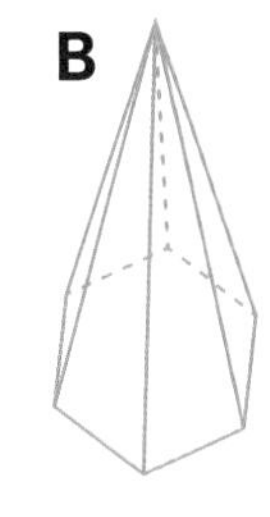

____

____

____

____

Challenge

1 Complete this table.

	3D shape	How many faces?	How many edges?	How many vertices?	How many bases?	Shape of base	Shape of side face
a							
b							
c							
d							
e							
f							

2 **a** Which object does this net make?

b Why would this net **not** make a triangular pyramid?

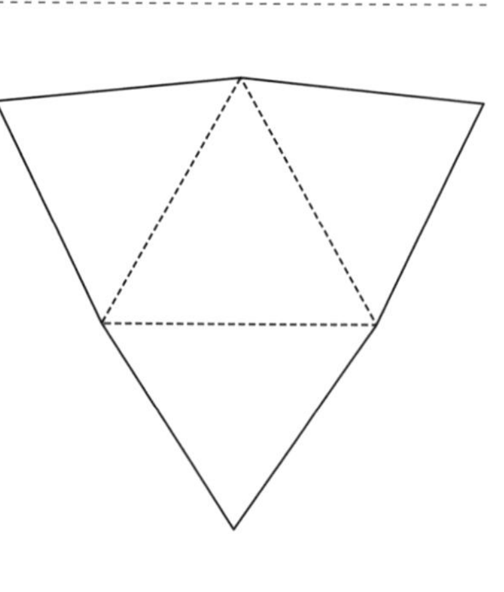

Mastery

1. In the student book you had the opportunity to copy some 3D shapes. Use the grid to practise drawing more 3D shapes.

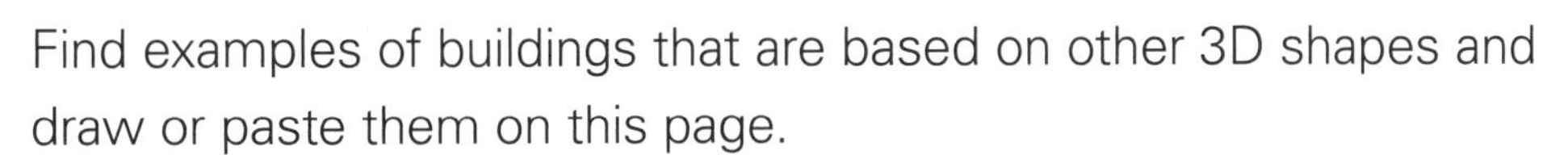

2. Perhaps the building you are in is based on a square or rectangular prism. This is the case with many buildings, because humans seem to love right angles. However, this first photograph shows a building in Germany that is based on four cylinders. The second photograph shows an octagonal-shaped building in England.

Find examples of buildings that are based on other 3D shapes and draw or paste them on this page.

Angles

Practice

1 How do you know that this is an acute angle?

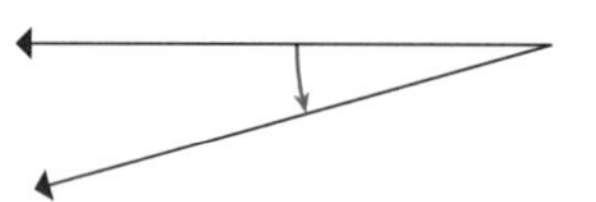

____

2 Write the type and size of each angle. (Be sure to read the correct protractor track.)

a An ______________ angle.

Size: ____°

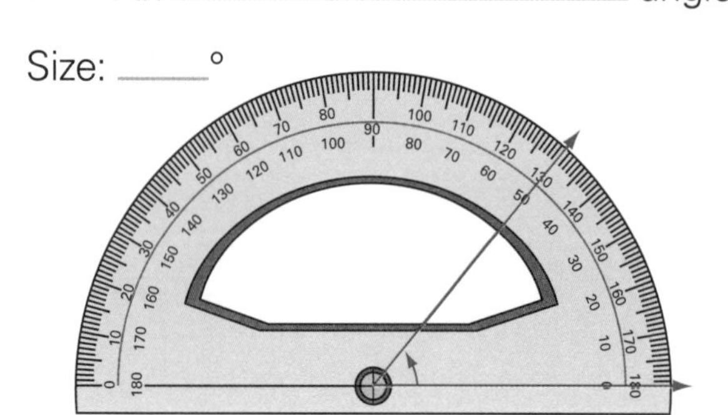

b An ______________ angle.

Size: ____°

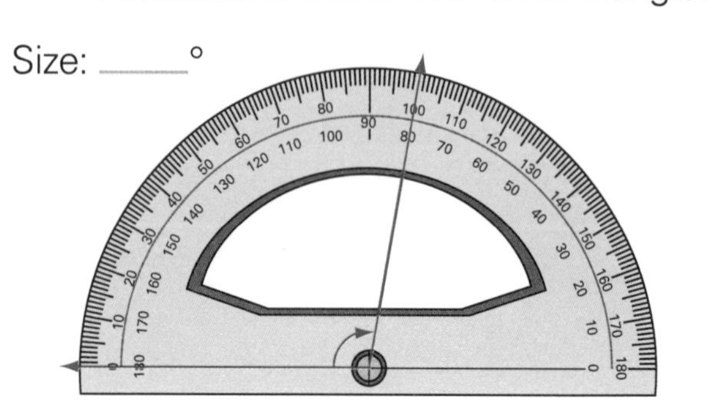

c An ______________ angle.

Size: ____°

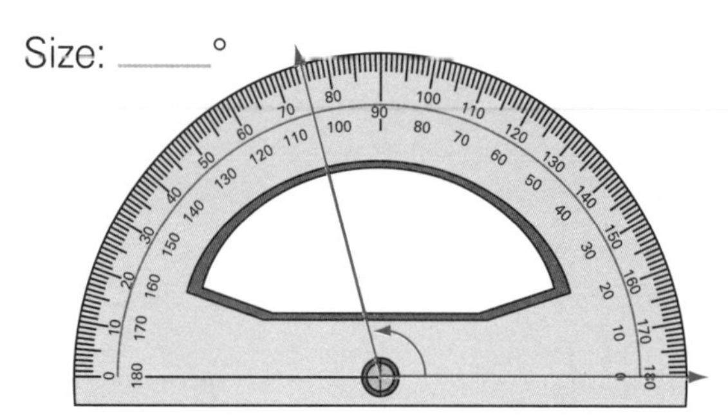

d An ______________ angle.

Size: ____°

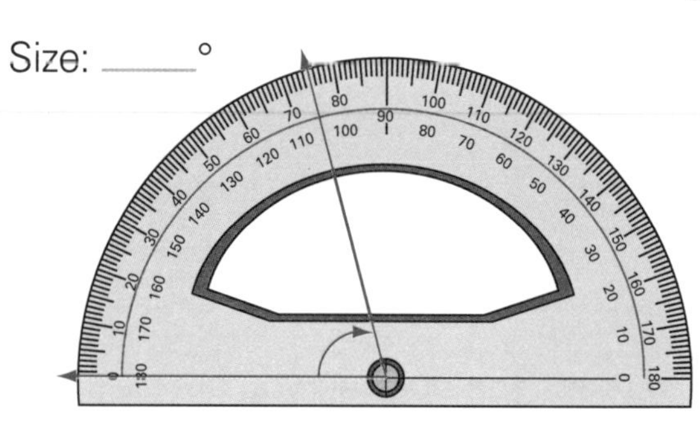

3 Write the type of angle. Circle the best estimate for the size of the angle.

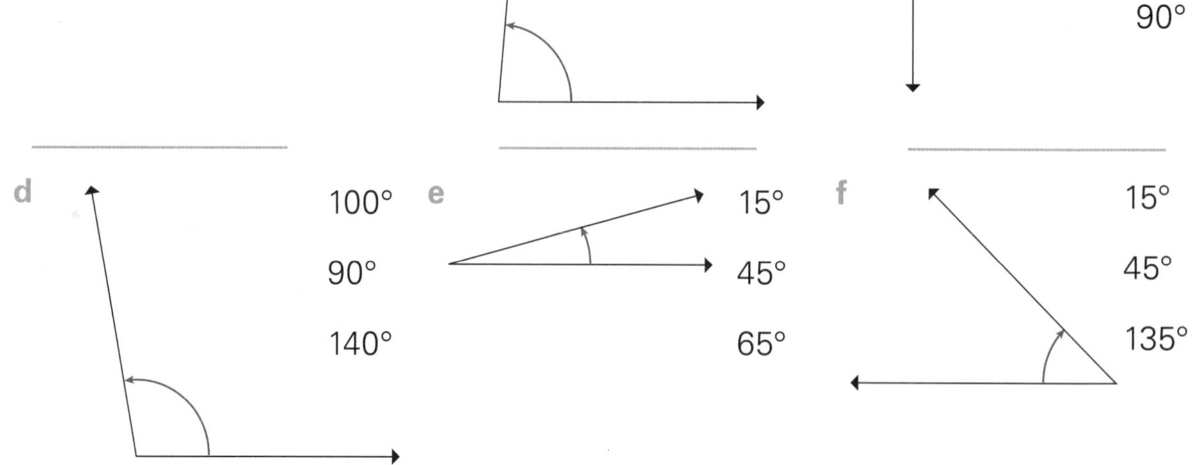

Challenge

1 Use a protractor to find the size of each angle. You may wish to make an estimate first.

a

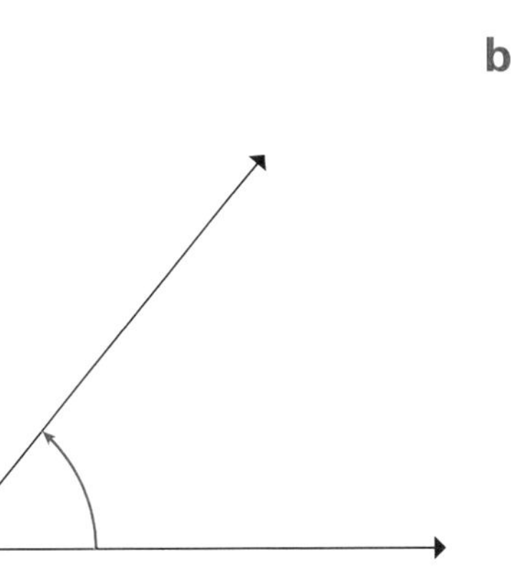

b

c

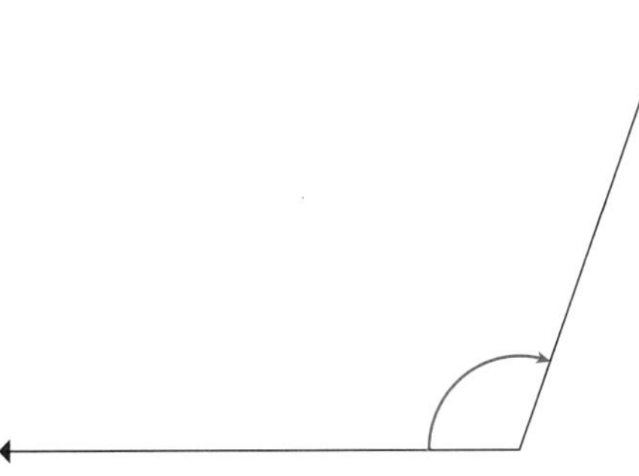

d

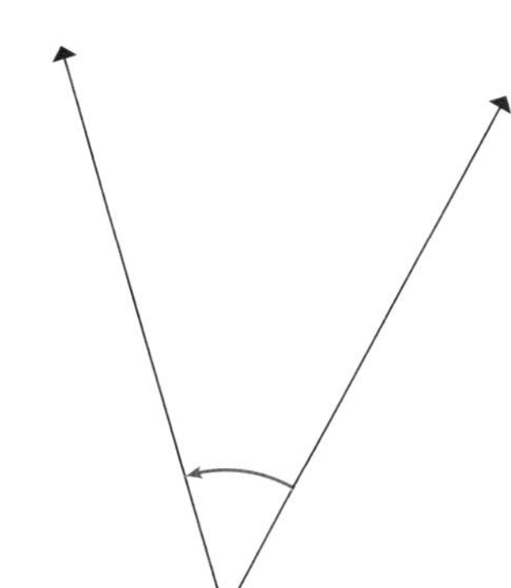

e

f

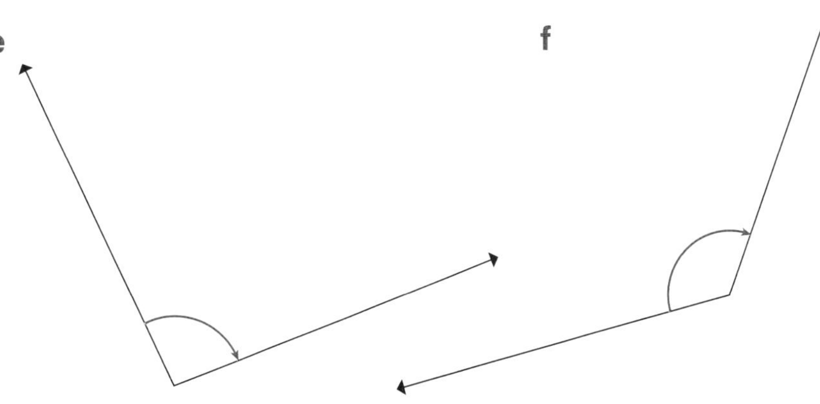

2 Use your choice of strategy to find the size of the two reflex angles.

a

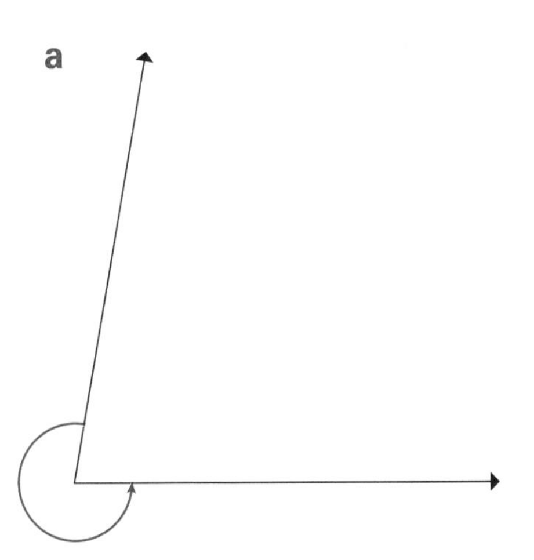

b

3 Use a protractor, pencil and ruler to draw the angle on each line. Start at the dot.

a 35°

b 165°

Mastery

1. Mathematicians say that no matter how a triangle is drawn, the sum of all three angles is always 180°.

   Find out whether this is true by measuring the angles in these triangles and finding the total of the angles for each.

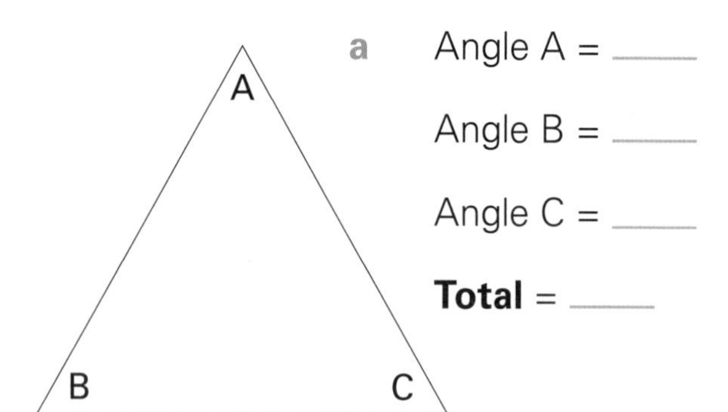

a Angle A = ____

Angle B = ____

Angle C = ____

Total = ____

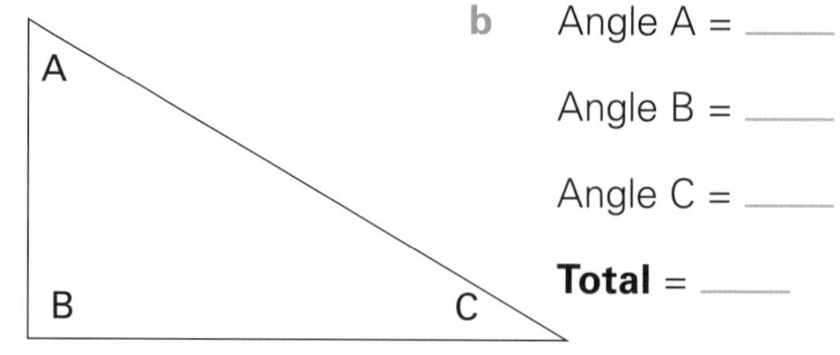

b Angle A = ____

Angle B = ____

Angle C = ____

Total = ____

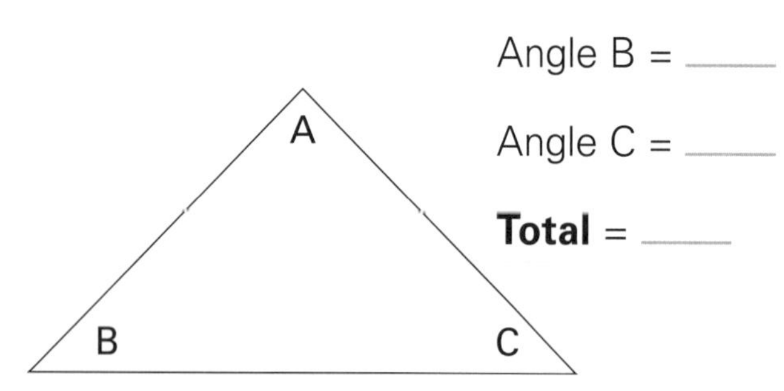

c Angle A = ____

Angle B = ____

Angle C = ____

Total = ____

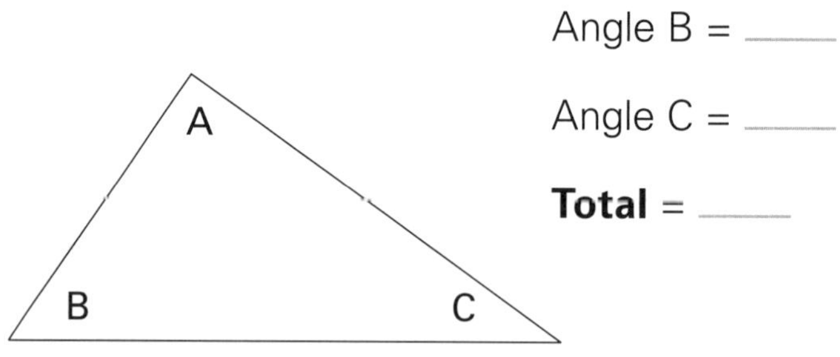

d Angle A = ____

Angle B = ____

Angle C = ____

Total = ____

2. Use the space to draw two triangles of your own. Measure each angle and then find the total. Comment on how close to 180° your totals are.

3. Find the sum of the angles of a rectangle. Use a separate piece of paper to find out whether the sum is the same for any quadrilateral.

Transformations

Practice

1 Have these shapes been transformed by translation, by reflection or by rotation?

a 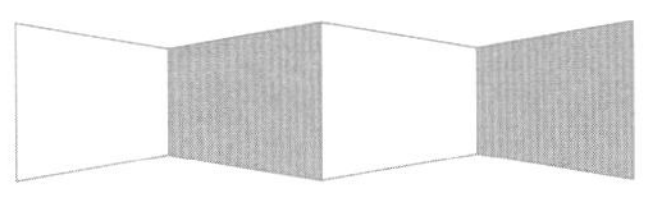

____

b 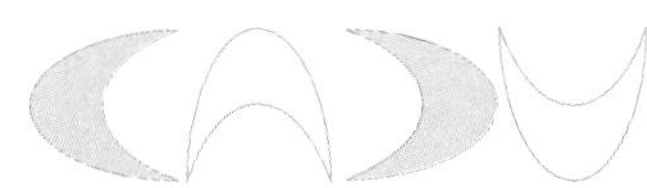

____

c 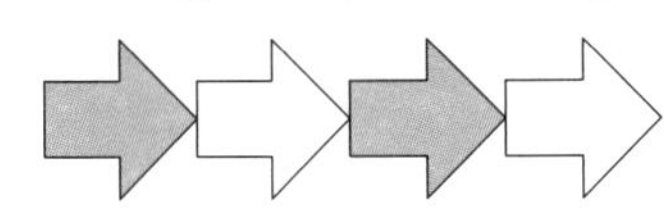

____

2 How have these patterns been made? Use the words **horizontally**, **vertically** or **diagonally** in your answers.

	Pattern	Description
a		The shape has been translated: ______________________
b		
c		
d		
e		
f		

3 Continue the pattern and describe the way it grows.

Challenge

1 Look carefully at how the pentagon has been transformed on each row. Continue the pattern over the whole grid. When you arrive at an edge, use only the part of the pentagon that will fit.

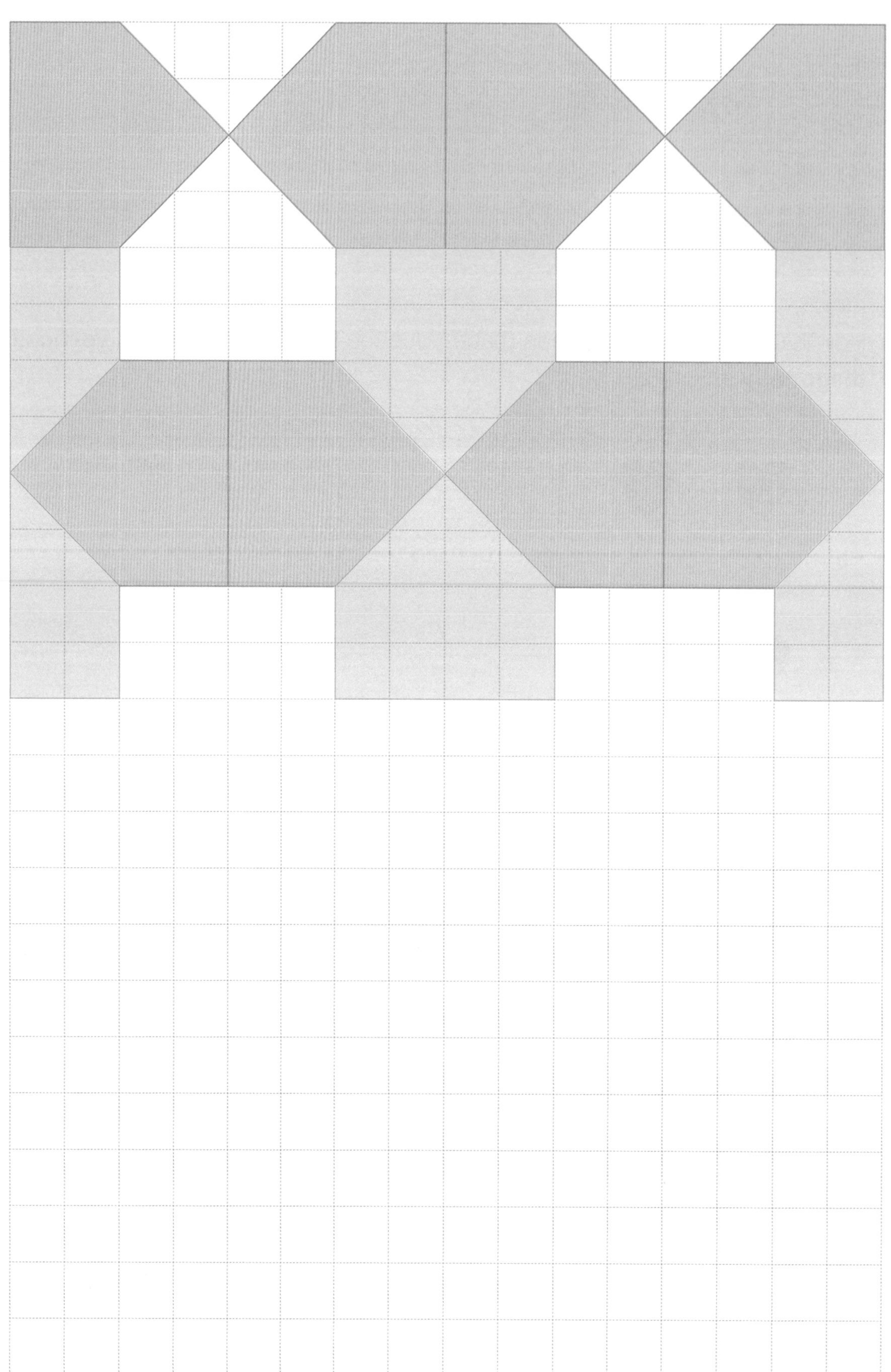

Mastery

1 On the last page you made a pattern by transforming a pentagon. People who design wall and floor tile patterns begin that task in a similar way. Choose a shape (or shapes) that you think would make an interesting transformation pattern. Draw it (and colour it if you wish) on this grid, just as you did with the pattern on the previous page. You may wish to begin by practising on spare grid paper.

UNIT 8: TOPIC 2
Symmetry

Practice

1. Look at the following shapes.

   a Put a small cross on the shapes that have no lines of symmetry.

   b Draw in all the lines of symmetry on the other shapes.

a

b 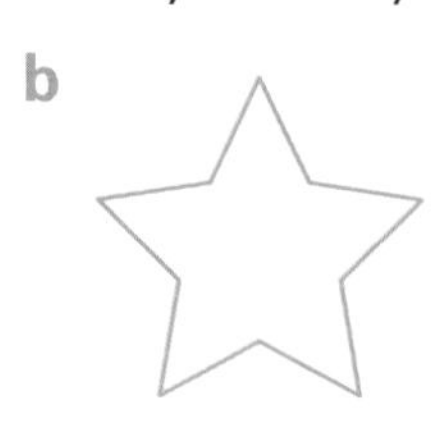

c 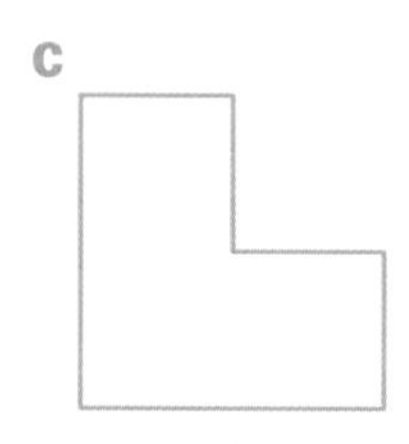

d 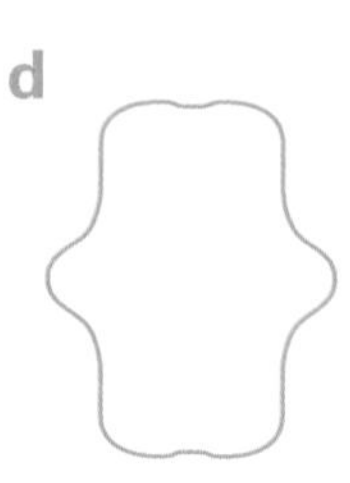

e

f 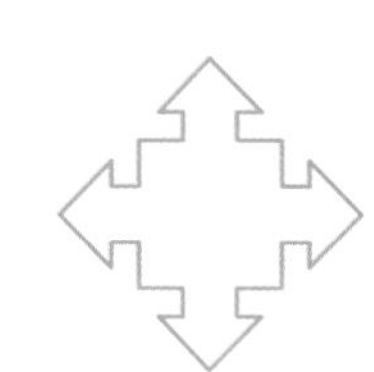

g 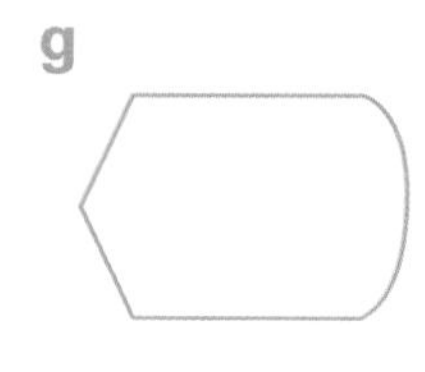

h 

i 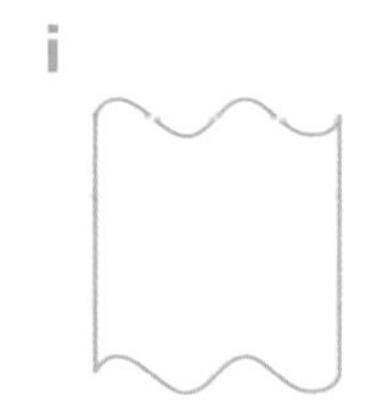

j 

k 

l 

2. Counting the starting position as one, write the order of rotational symmetry for each of the following shapes:

a 

Rotational symmetry of order ______

b 

Rotational symmetry of order ______

c 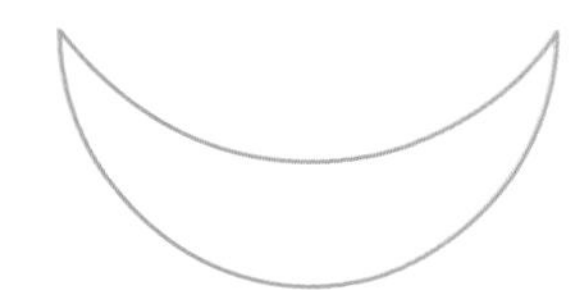

Rotational symmetry of order ______

d 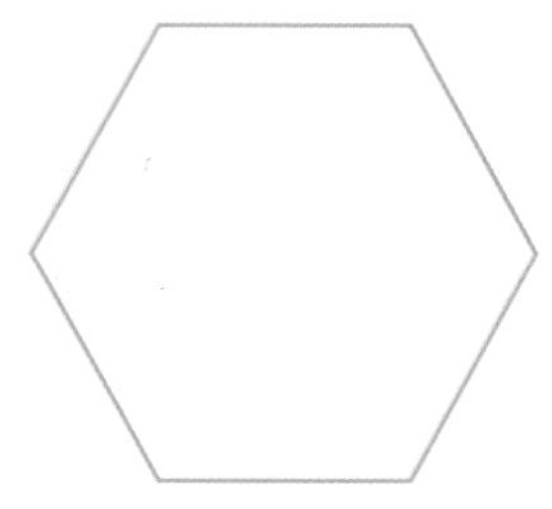

Rotational symmetry of order ______

e 

Rotational symmetry of order ______

f 

Rotational symmetry of order ______

Challenge

1 **a** Jack wants to write a word in which every letter has at least one line of symmetry. He decides on the word SHOE. Explain why that is not correct.

---

b Write a word (or words) in which all the letters have at least one line of symmetry. Show that you are correct by drawing in the lines of symmetry.

2 The word ICEBOX can be written with a horizontal line of symmetry. **ICEBOX**

Find words with horizontal lines of symmetry. Show that you are correct by drawing in the lines of symmetry.

3 The word DAD does not have a vertical line of symmetry, but the word MUM does.

Find words in which the whole word has a vertical line of symmetry. **MUM**

Show that you are correct by drawing in the lines of symmetry.

4 This is half of a symmetrical picture. Use the grid to complete the picture.

Mastery

1. The Australian flag has some parts that are symmetrical, such as the stars.

The Union Jack in the top left-hand corner looks as though it has some lines of symmetry but, if you look closely, you will see that it is asymmetrical (not symmetrical). What is it that makes it asymmetrical?

____

2. The Canadian flag has a line of symmetry through the maple leaf in the centre.

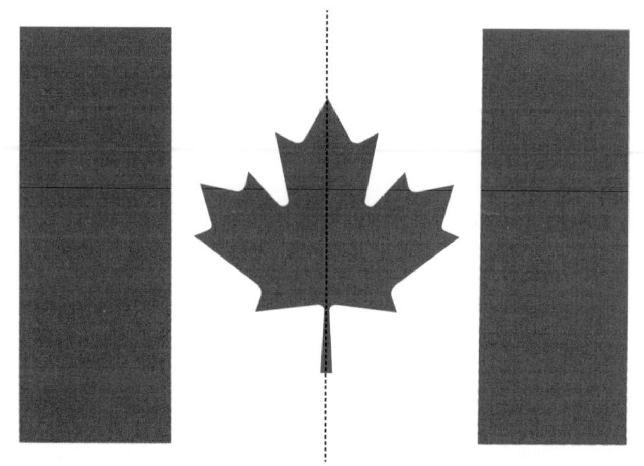

Carry out some research to find flags that are symmetrical. Draw them, and include some information about the flags. Draw in the line(s) of symmetry. You could make your work more interesting by including symmetrical flags that are made up of designs other than just stripes.

UNIT 8: TOPIC 3
Enlargements

Practice

1 Make an enlargement of these shapes on the second grid. Then make an even bigger enlargement by drawing them onto the third grid.

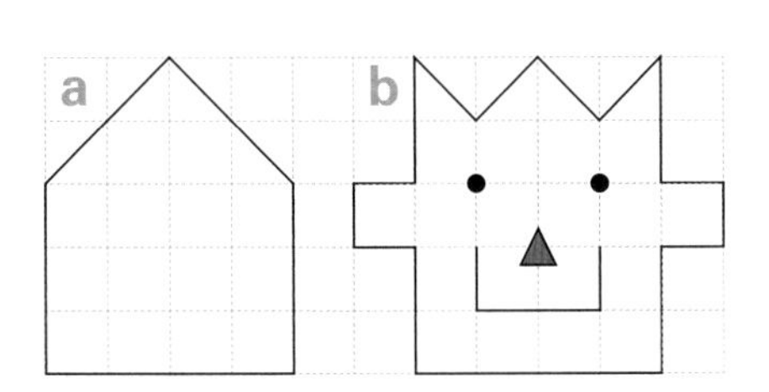

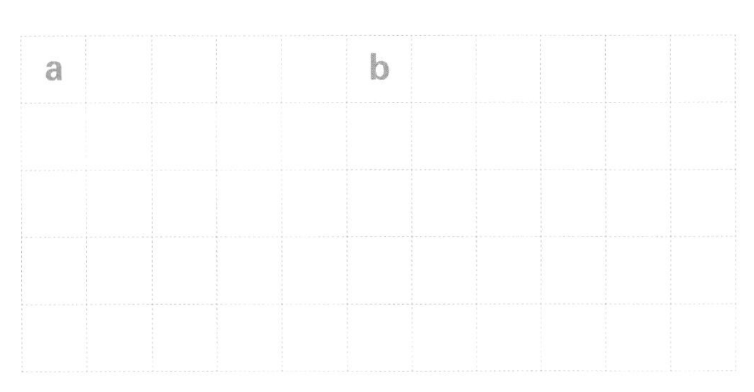

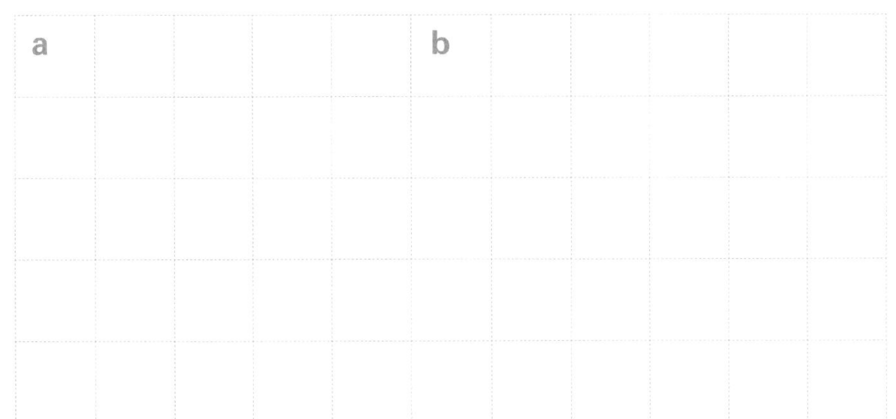

2 Make a reduction and an enlargement of the letters by drawing the word on a smaller and a larger grid.

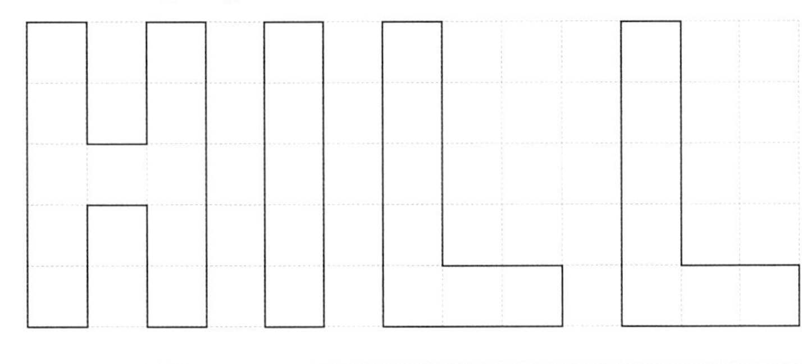

Challenge

1 Redraw the pictures according to the scale factors shown. Start each drawing at the red dot.

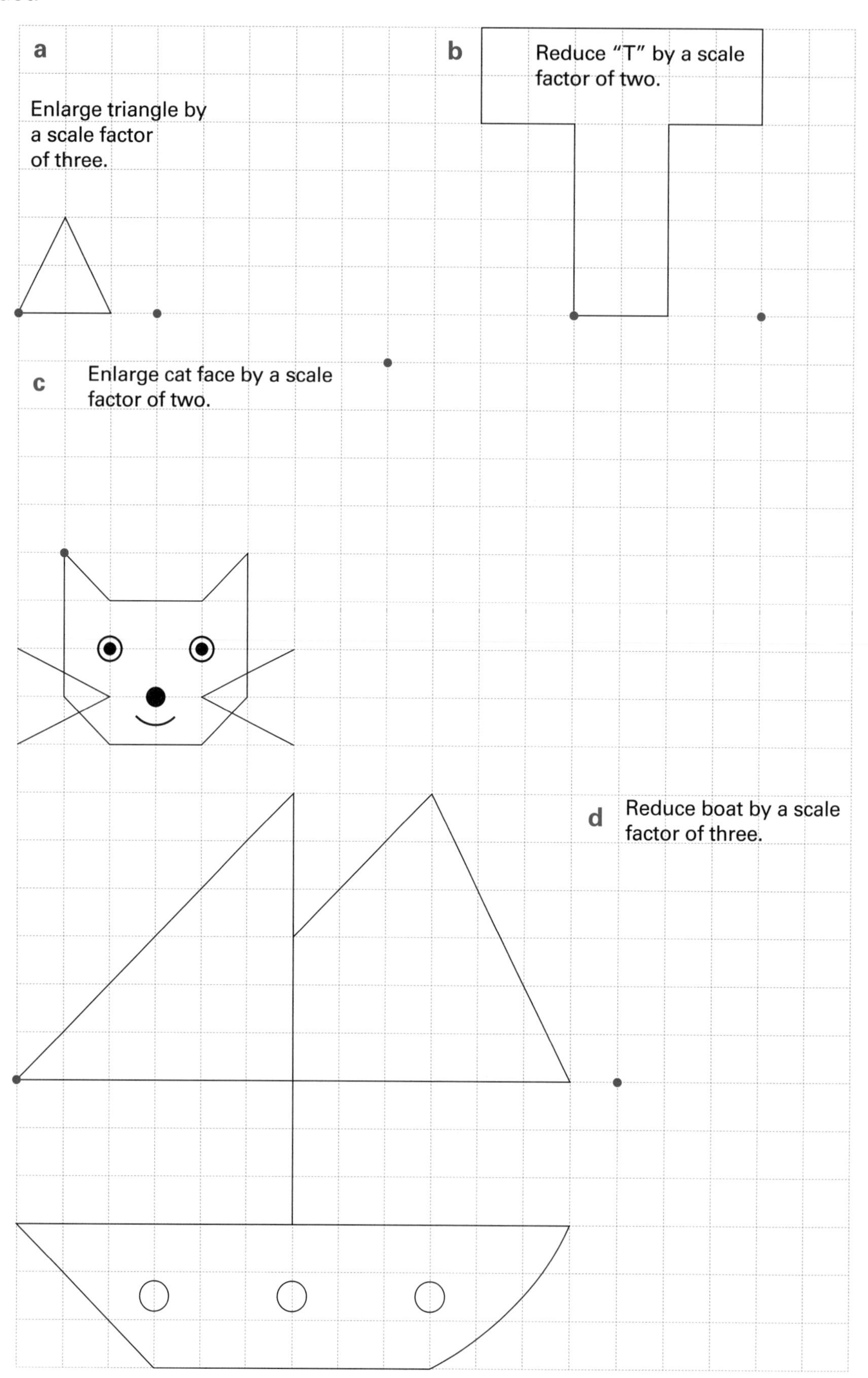

Mastery

1. In this topic you have enlarged and reduced drawings by a scale factor of two or three. These are not the only scale factors that can be used. How small you can make the drawing depends on patience and accuracy. How big you can make it depends on the size of paper you can get. In theory it would be possible to enlarge a 5-centimetre picture onto the side of a house!

   Find a picture to use that you find interesting. It could be a photo of you, a face from a magazine, or a picture by a famous artist. You will probably need spare grid paper for this activity. You may wish to begin with a simple design, such as this house shape. Try enlarging your picture by a scale factor of four or five.

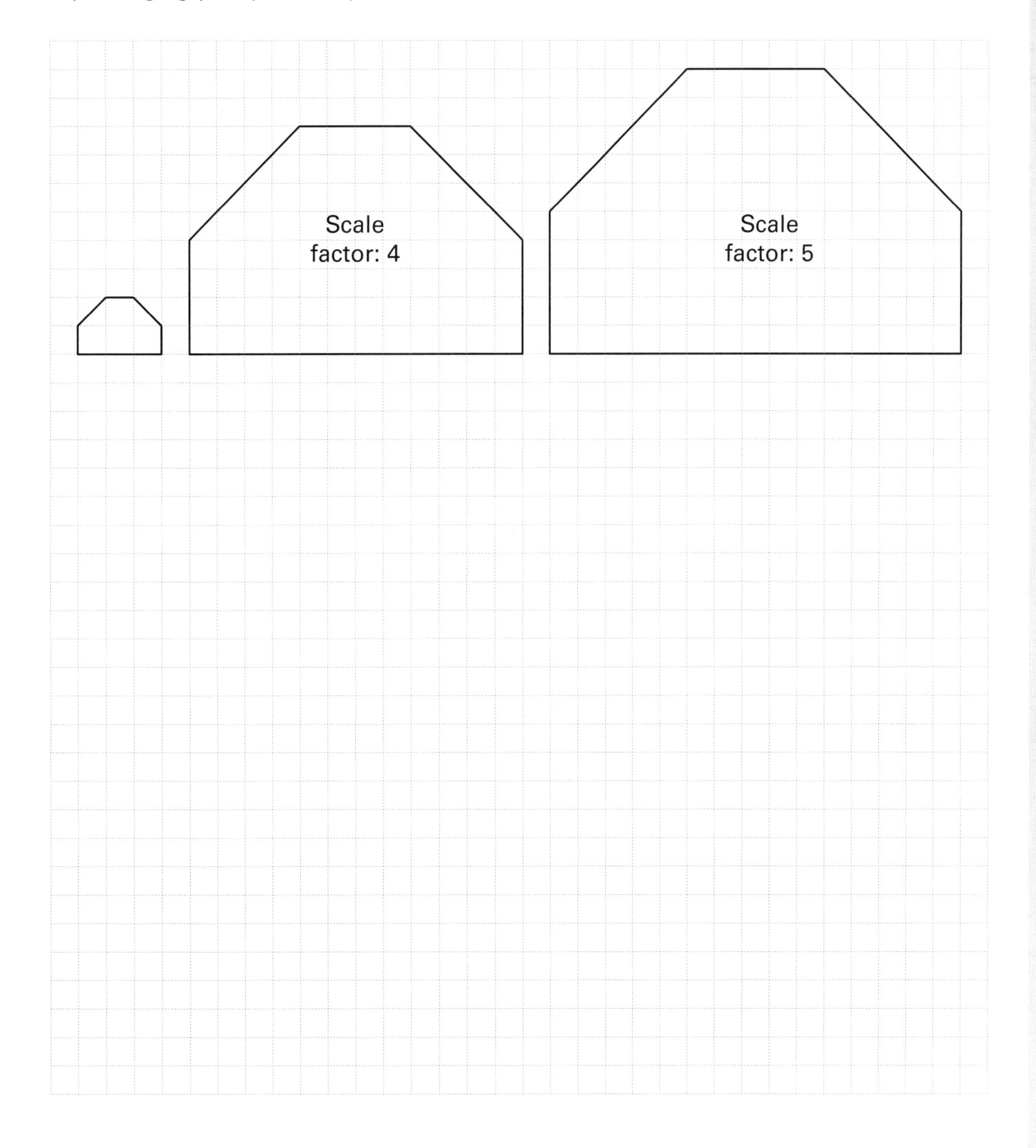

UNIT 8: TOPIC 4
Grid references

Practice

The same shapes have been drawn on these three grids, but the grids are not all the same. Look carefully at the grids to complete the activities. Remember to go across the *x*-axis first, then up the *y*-axis.

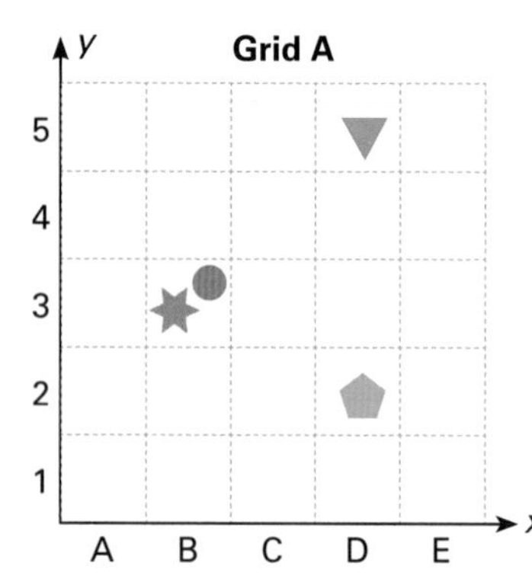

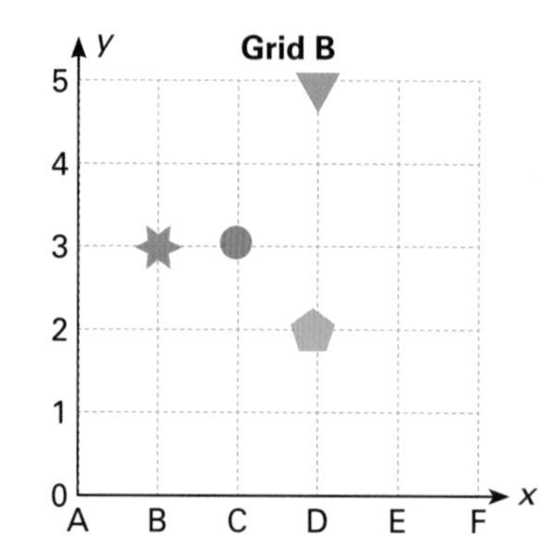

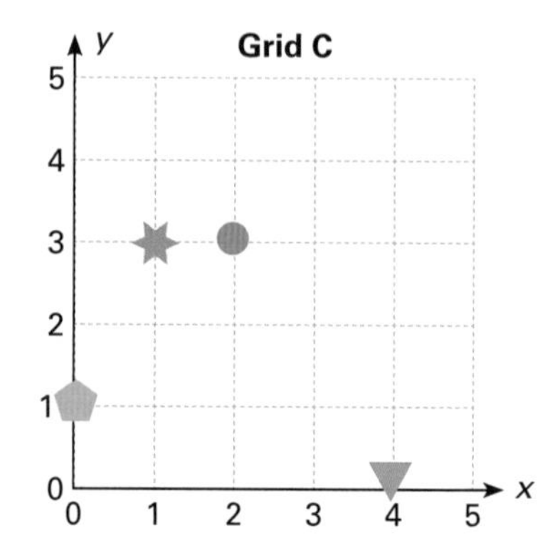

1 Write the grid reference or coordinate point for the position of the circle on:

a Grid A: ________ b Grid B: ________ c Grid C: ________

2 It is only possible for the circle and the star to have the same grid reference on one of the grids. Which grid? Explain why.

____

3 Write the grid reference or coordinate point for the position of the triangle on:

a Grid A: ________ b Grid B: ________ c Grid C: ________

4 Write the grid reference or coordinate point for the position of the pentagon on:

a Grid A: ________ b Grid B: ________ c Grid C: ________

5 a Draw a small cross on the following coordinate points on Grid C: (3,1) (3,4) (5,4)

b Write a coordinate point that would make the crosses form a rectangle. Draw a cross at that point. ________

6

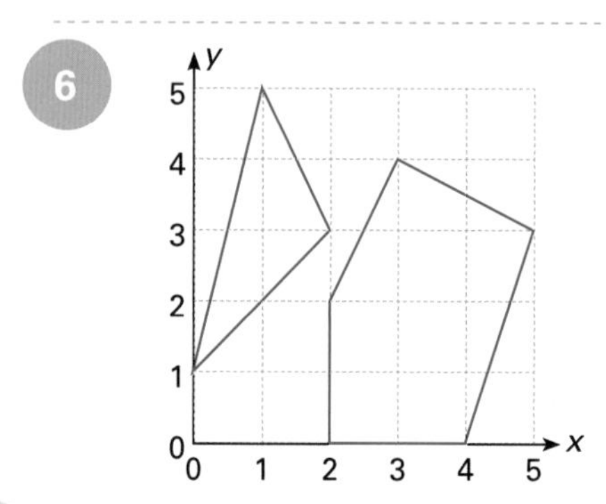

a The coordinate points for drawing the triangle are (1,5) ⟷ ________ ⟷ ________ ⟷ ________

b The coordinate points for drawing the pentagon are (3,4) ⟷ ________ ⟷ ________ ⟷ ________ ⟷ ________ ⟷ ________

Challenge

1 **a** Draw a capital W and a capital M on the grid as follows:

- Each letter is 3 squares high and 4 squares wide.
- The W starts at (1,6).
- The M starts at (1,0).

b Write the coordinate points to show someone how to draw the following letters:

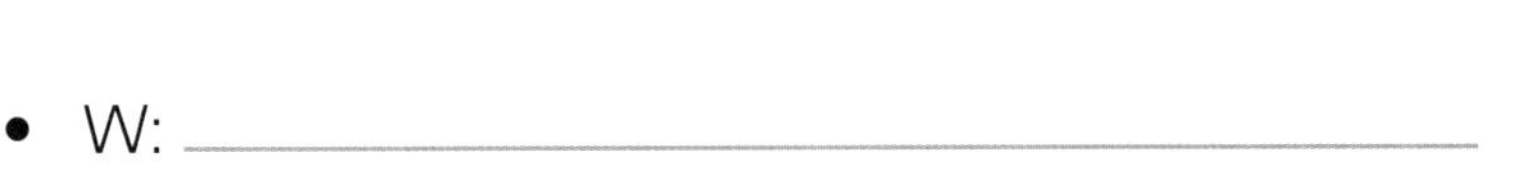

- W: ________________________________
- M: ________________________________

---

2 This is half of a symmetrical drawing for a simple picture of outer space:

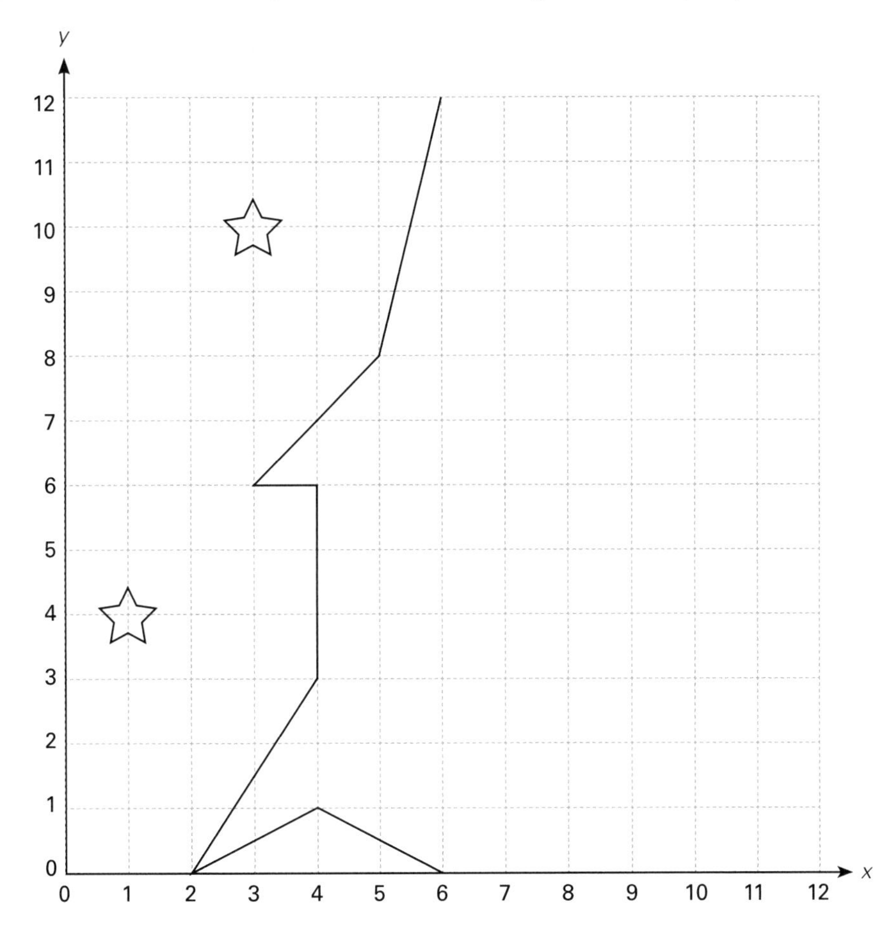

a Draw the other half of the picture, making sure that it is symmetrical.

b Starting at (6,12), write instructions to draw the rocket, just as you did for the letters in question 1.

____

____

c Write the four coordinate points for the stars.

____

Mastery

1 It is easier to find your way around places like theme parks if you have a map. The places on the map are usually shown inside the squares on a grid.

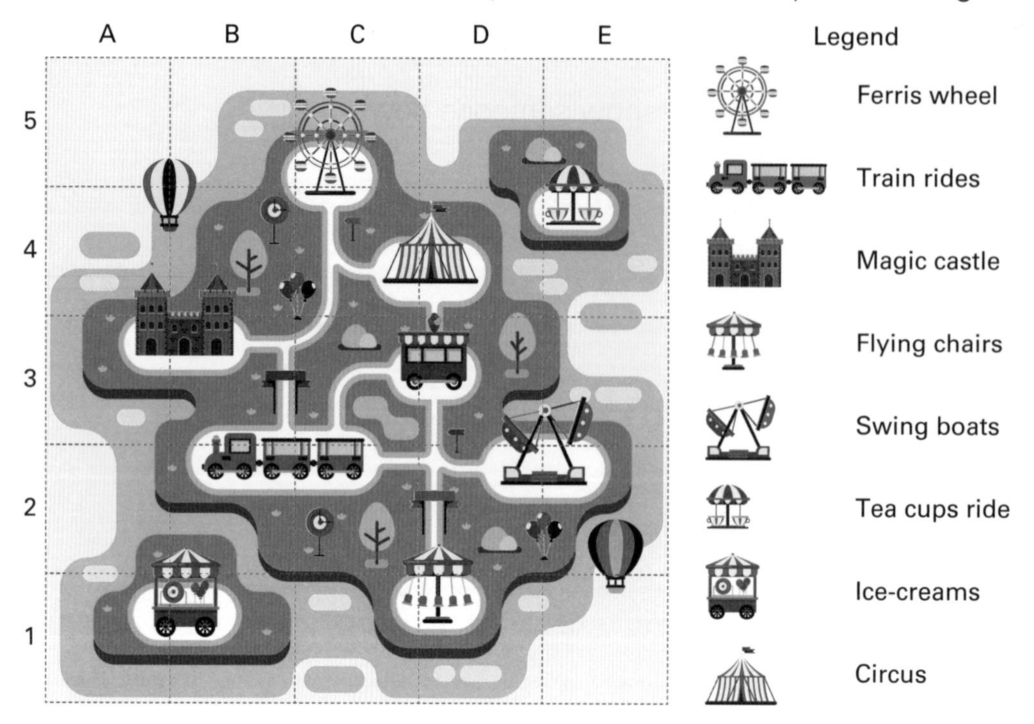

Use the map and the legend to complete these activities.

a Give the grid reference for the ferris wheel. ________

b In which four squares is the castle? ________________

c Write a question for which the answer could be E4 and E5. ________________

---

2 Draw a map on the grid to design your own park. Choose a theme and include a legend. When you are finished, write some tasks based on your map.

Giving directions

Practice

1

a On the compass rose, Tom is to the south of Sam. Put "S" for south on the compass rose.

b Label the other seven points using "N" for north, "W" for west, and so on.

c Draw a smiley face north-east of Sam.

d Who is south-west of Sam? ________

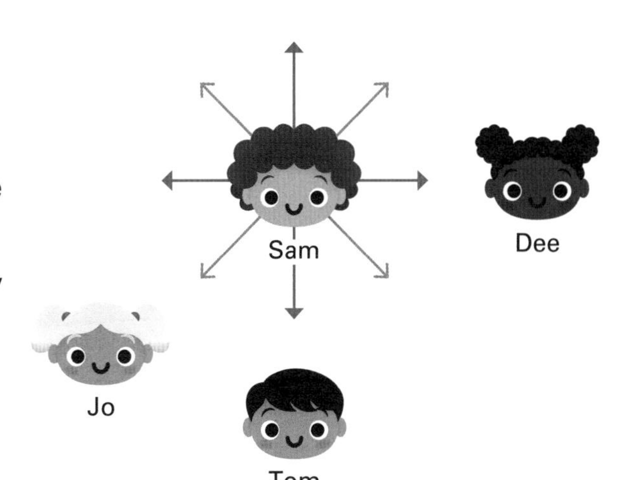

2 Use the map to complete the activities below.

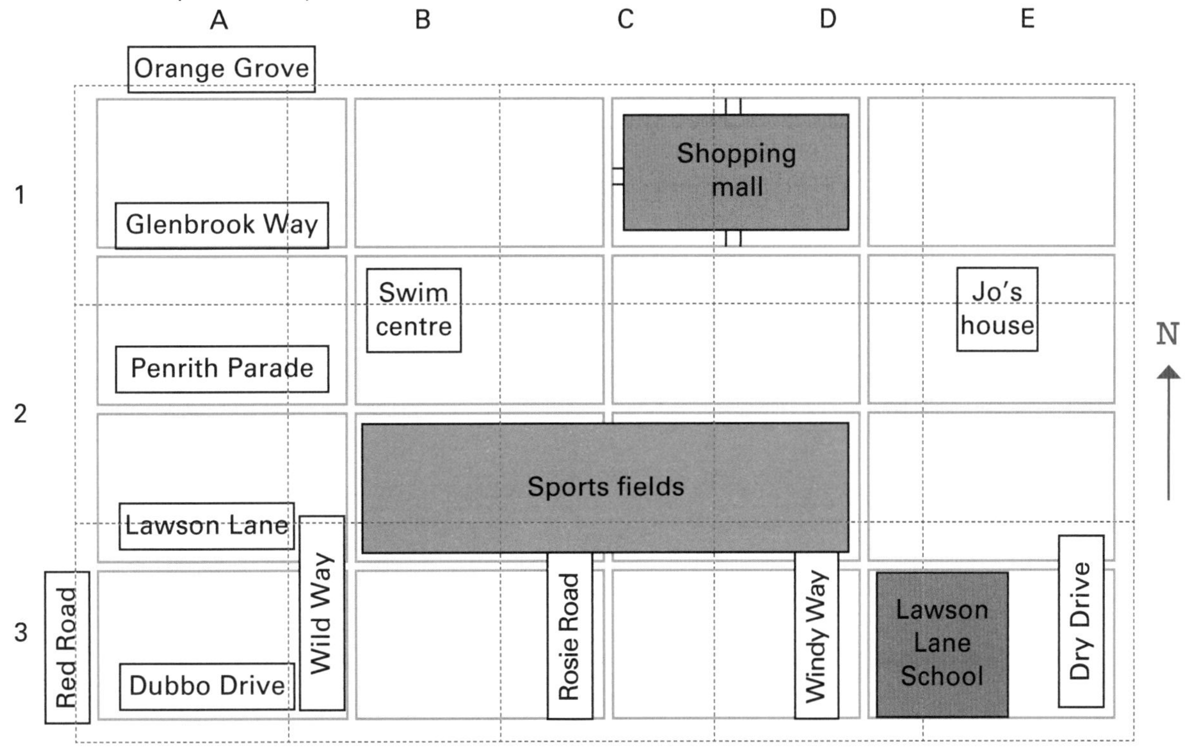

a In which direction would Jo travel to get from her house to the swim centre? ________

b Which direction is the school in E3 from the sports fields in C2? ________

c Imagine you take the south exit from the sports fields and go west along Lawson Lane. You turn north at the next corner. What is on your right just before Glenbrook Way? ________

d There's a new cinema in town. It's so new, it's not even on the map! The cinema is north-west of the sports fields in B2. It's on the southern side of Glenbrook Way, near the junction with Red Road. Draw and label it with a "C" on the map.

Challenge

This is a map of Deeton:

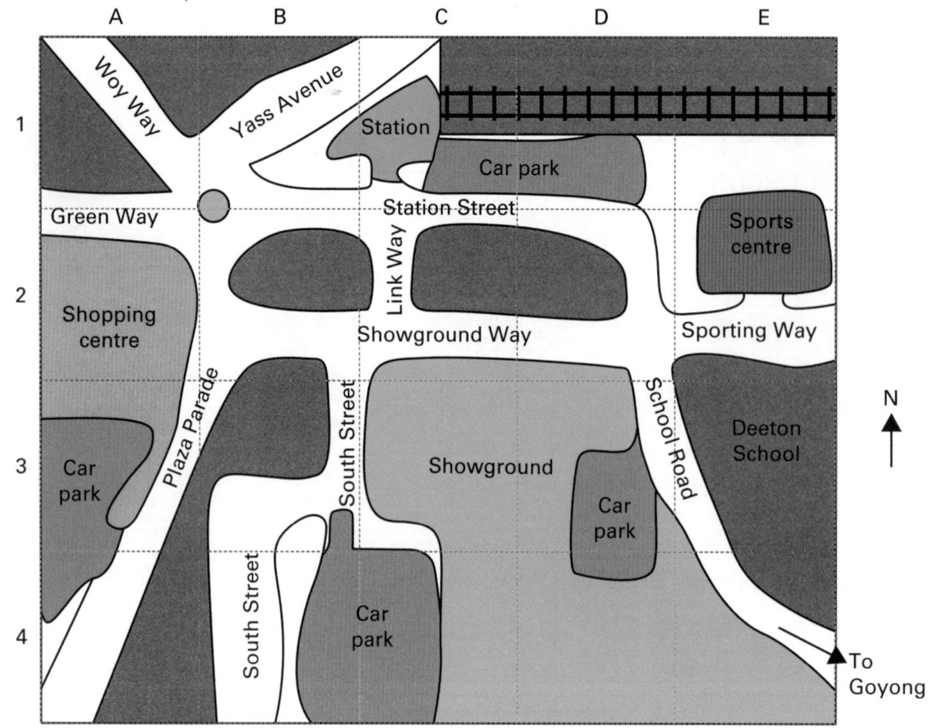

1 Give grid references for:

a the station ________ **b** the sports centre ________

c Deeton School ________

2 Name the roads that pass through:

a C1 ________

b C2 ________

3 **a** Which roads run in an east–west direction? ________

b Which road goes north-west from the roundabout? ________

4 Which direction is it from Deeton to:

a Goyong? ________ **b** leave town on a train? ________

5 Using street names and compass directions, give directions to go from:

a the sports centre to the shopping centre

____

b the car park exit in B3 to the shopping centre entrance in A2

____

Mastery

1 The town of Deeton is an imaginary place. The artist who drew the map had to decide which roads and map features would be in this imaginary town. This is actually similar to the way many new towns and suburbs are planned today.

Carry out some research to find out about a planned town, suburb or city near you, and write a few sentences about the person or people who planned it.

2 In this task it is your turn to draw a map. You could invent a place that is just the way you want it to be. You could draw a map of your school, a "treasure island" or the town where you live. You could use this grid for the map so that it is easy to locate various places on the map, remembering to put letters/numbers along the two axes of the grid.

Collecting and representing data

Practice

Remember: Numerical data can be counted or measured.

1. Write a survey question about books that would enable you to collect:

   a numerical data

   ____

   b categorical data

   ____

2. Create a two-way frequency table about the colour of 20 people's eyes. Show whether the information is from a male or a female.

Eye colour M or F	Blue	Brown	Other	Total
Female				
Male				
Total				

3. Use the data about eye colour to create a bar graph and a dot plot. Decide on a suitable scale for the bar graph.

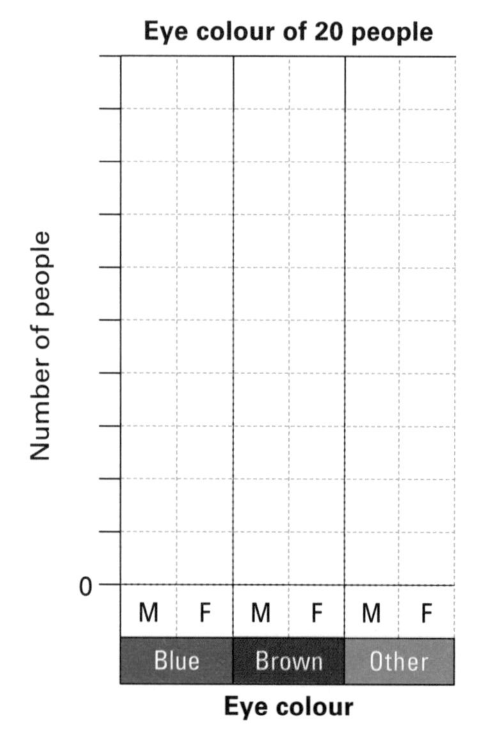

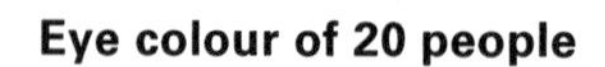

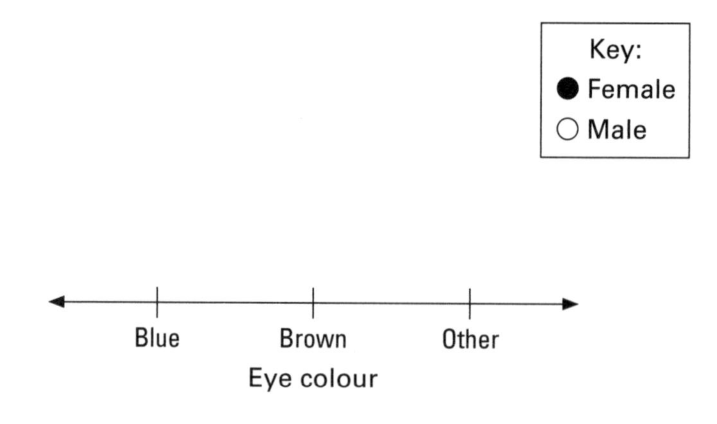

Challenge

1 Earth turns on its axis once every day. The length of a day on Earth is not the same as it is on other planets in the Solar System. The planet Mercury turns very slowly. If you could spend a day on Mercury and watch the sun set, it would be about three Earth months before you saw it rise again!

This circle graph compares the length of a day on six of the planets in our Solar System.

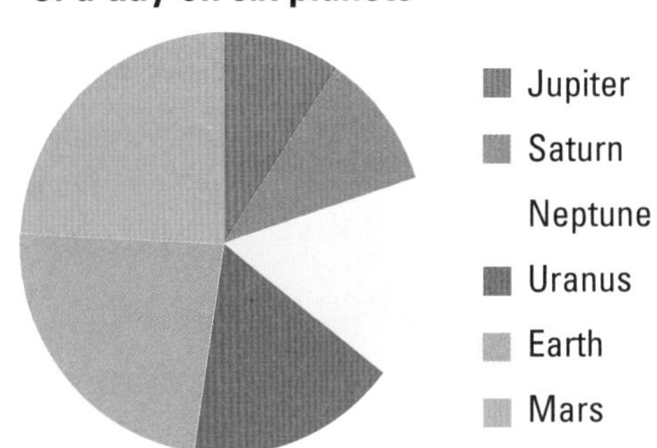

It is not possible from the circle graph to see exactly how long a day is on each of the planets, but we can estimate their lengths by comparing them with an Earth day.

a How many hours are in an Earth day? ______

b Which planet has a day that is almost the same length as an Earth day? ______

c Which two planets take the least amount of time to turn once? ______

2 It is possible to compare the length of the days on the six planets more accurately.

a This table shows the time it takes for each planet to turn once. Round the times to the nearest whole hour.

Planet	Amount of time it takes to turn once	Amount of time it takes to turn once, rounded
Earth	23 hours 56 minutes	
Mars	24 hours 40 minutes	
Jupiter	9 hours 54 minutes	
Saturn	10 hours 40 minutes	
Uranus	17 hours 14 minutes	
Neptune	16 hours 11 minutes	

b Transfer the data to a horizontal bar graph. Decide on a suitable scale. Give the graph a title and label it correctly.

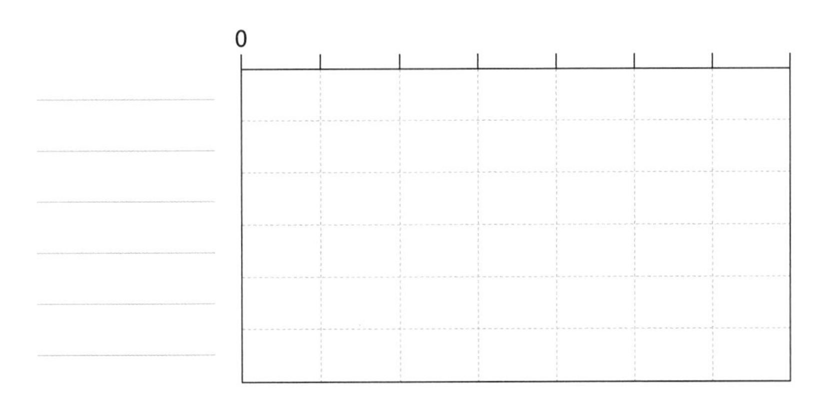

Mastery

1 TV networks want to make sure that people will watch their programmes. They carry out lots of research to find out about people's TV viewing preferences.

Imagine that you work for a TV company. Collect and present some **numerical** and some **categorical** data that would be useful for the company. Here are some starting points that you could use:

- When do people watch TV?
- How long do they watch for?
- What sort of programmes do they like?
- What age group do they fit into?

Decide on a suitable survey question for each type of data. Carry out the research, and decide on a suitable way to present each set of data. Then create the graphs.

You could use the grid below, or you may prefer to use some separate paper.

UNIT 9: TOPIC 2
Representing and interpreting data

Practice

1 Audrey played 10 games of cricket for her school. These are her scores. Represent the data on a line graph.

Game	1	2	3	4	5	6	7	8	9	10
Score	5	10	5	0	25	35	49	31	0	50

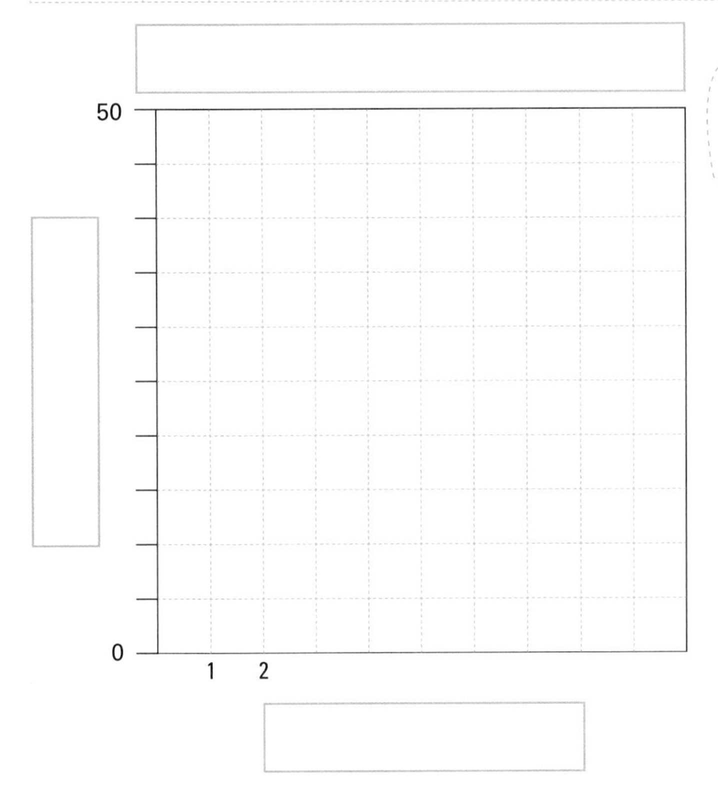

A line graph shows data that changes over time.

a Complete the numbering for the vertical axis and the horizontal axis.

b Write a title for the graph.

c Label the horizontal and vertical axes appropriately.

d Plot the data and join up each point

2 a Between which games did Audrey's score increase by 10 runs?

____

b Describe the change in scores between games 7 and 9. _______________

c In which two games did Audrey get her best scores? _______________

d What was Audrey's total score in the 10 games? _______________

e In which games did Audrey score **less** than her average score?

____

Challenge

1 Car drivers need to be ready to stop at any time. However, a car cannot stop immediately. How long it takes to stop depends on three main things: the **thinking distance**, the **braking distance** and the **speed** the car is going.

- The **thinking distance** depends on the time it takes for a driver to think, "I need to stop," and then for the brain to tell the foot to press the brake pedal.
- The **braking distance** is how far the car travels after the brake pedal has been pressed.
- The **speed** is important because the faster the car is going, the further the car travels during **thinking** and **braking**.

This table shows how many metres a car travels during the **thinking** and **braking** times at various speeds.

Speed	25 km per hour	50 km per hour	75 km per hour	90 km per hour	110 km per hour
Thinking distance	5 m	15 m	30 m	45 m	
Braking distance	5 m	20 m	45 m	65 m	
Total stopping distance					

a Decide on a likely thinking and braking distance for a car travelling at 110 km per hour and write them in the table.

b Fill in the totals for a car's stopping distance for each speed.

2 Use the data from the table to create a horizontal bar graph.

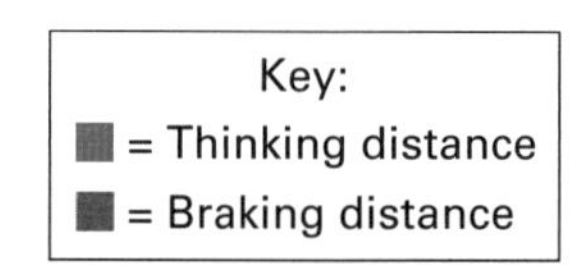

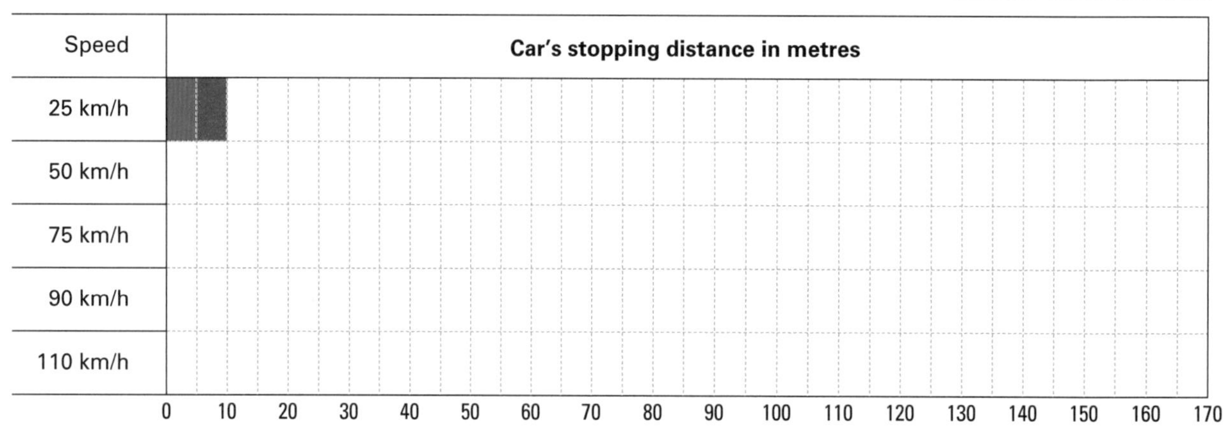

3 What do you notice about the differences between the thinking and braking distances once a car travels at over 25 km/h?

Mastery

1. Perhaps you are taller, smaller, or about the same as the average height for people who are your age.

   To find the average height for a person your age, you need to collect some data.

   a Decide how many people you need to measure to find the average height for your age.

   b With your teacher's permission, collect the information.

   c Decide on a suitable type of graph. You may wish to use the grid below or a separate piece of paper to construct your graph.

   d Decide on a range for the scale of your graph and label it.

   e Represent the data on your graph. Don't forget to make it look neat.

   f Calculate the average height.

   g Draw a line across the graph to show the average height of a person your age.

UNIT 10: TOPIC 1
Chance

Practice

1 The following sentences contain chance words to describe the likelihood of something occurring.

Fill in the gaps by converting these chance words to a decimal, a fraction and a percentage to describe the chance of each event occurring.

	Description	Value: Decimal	Value: Fraction	Value: %
A	It is *almost certain* that someone will laugh today.		$\frac{9}{10}$	
B	It is *certain* that someone will talk today.			
C	It is *very unlikely* that I will see a movie star this weekend.			
D	It is *unlikely* that I will like every lesson in high school.			
E	It is *very likely* that I will make a spelling mistake this year.			
F	There is *a better than even chance* that someone will sing today.			
G	They say there is *a less than even chance* that it will rain.	0.4		
H	It is *impossible* that:			
I	It is *almost impossible* that:			
J	It is *likely* that I will smile today.			
K	There is an *even chance* that:			50%

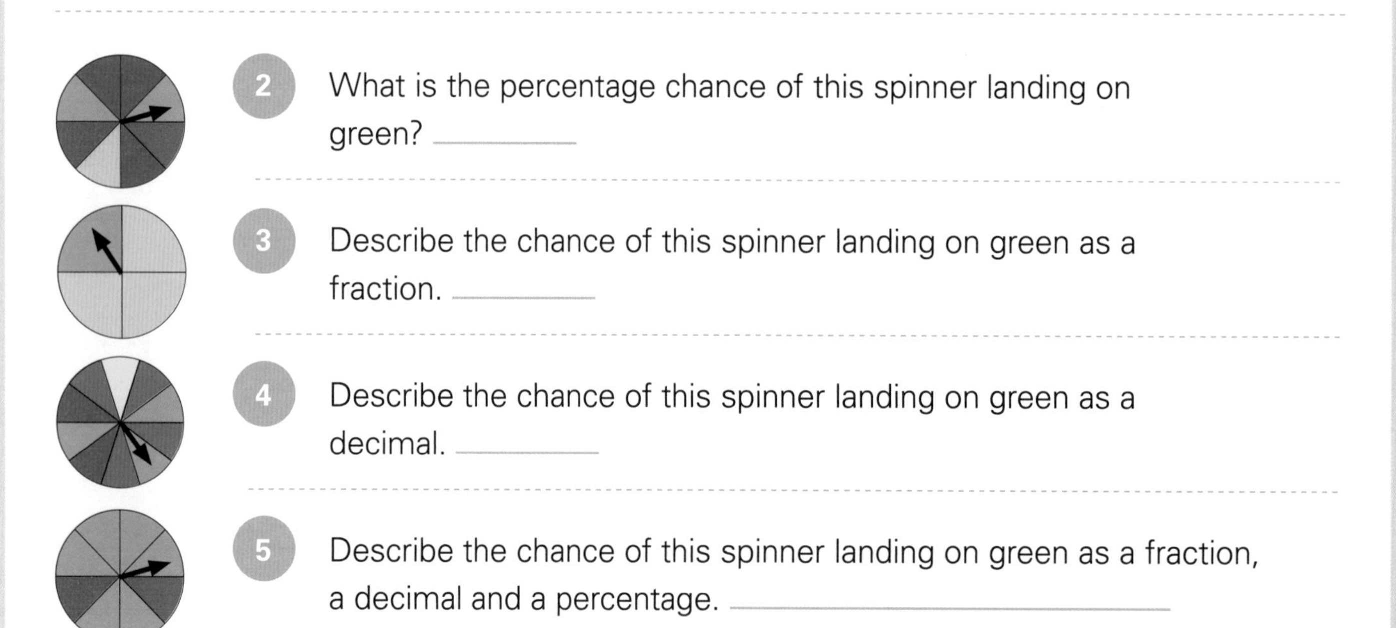

2 What is the percentage chance of this spinner landing on green? ________

3 Describe the chance of this spinner landing on green as a fraction. ________

4 Describe the chance of this spinner landing on green as a decimal. ________

5 Describe the chance of this spinner landing on green as a fraction, a decimal and a percentage. ________

Challenge

1 Only two of the 20 sections of this spinner are green.

Based on how the spinner is coloured at the moment, write the chance of the spinner landing on green as:

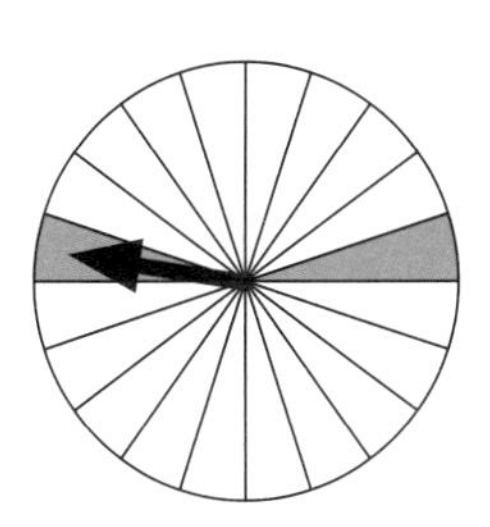

a a fraction ________ **b** a decimal ________ **c** a percentage ________

2 **a** Colour the other sections of the spinner so that the following probabilities are true:

- There is $\frac{1}{10}$ chance for red.
- There is a 0.3 chance for yellow.
- Blue has a 25% chance.
- There is $\frac{1}{5}$ chance for purple.

b The remaining sections are left white. Write the chance of the spinner landing on white as:

- a fraction ________
- a decimal ________
- a percentage ________

3 There are 48 beads in each cup, some blue and some white.

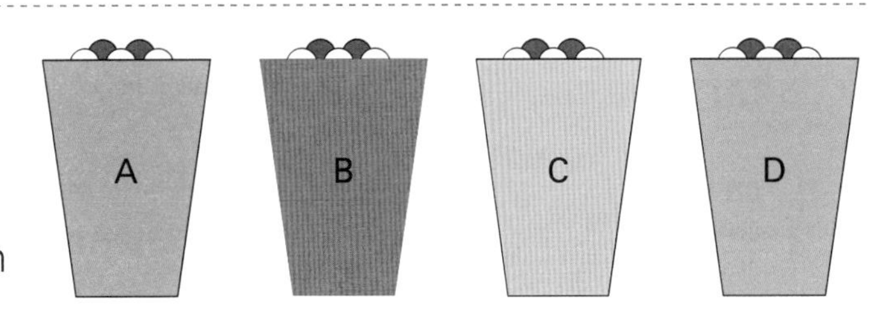

Without looking, Sam takes eight beads from each cup and writes down the number of each colour. Using chance values, predict how many blue and white there will be when all 48 beads have been taken out.

	After 8 have been taken out		My prediction after all 48 have been taken out	
Cup A	blue: 2	white: 6	blue:	white:
Cup B	blue: 4	white: 4	blue:	white:
Cup C	blue: 5	white: 3	blue:	white:
Cup D	blue: 7	white: 1	blue:	white:

4 Evie decides to roll a dice six times. She wants it to land on six on the first roll. It does!

a Describe the chance of a six-sided dice landing on six as a fraction. ________

b Evie rolls the dice another four times. To her amazement, it lands on six every time. What is the chance that it will land on six on the final roll? ________

Mastery

1 Tanya used a spinner in an experiment. This table shows the results:

Outcome	Red	Blue	Yellow	Green
Number of times spinner landed on each colour	7	53	27	13

a Draw a picture of what you think the spinner looked like.

b Give an explanation for the way you coloured the spinner.

____

____

____

____

2 Most board games depend on chance, and that's what makes them fun. In the game of Monopoly, for example, whether you win or lose depends mostly on the way the dice land.

In this task you have to invent a board game that depends on **skill** as well as **chance**.

For example, if you had a hundred grid and a pair of dice, the winner might be the one who gets to 100 first, but that would depend entirely on chance.

However, the game could be adapted so that some squares have questions. If a question is answered correctly, the player moves forward 10 squares; a wrong answer could mean going 10 squares back.

You can probably think of a much better **skill** and **chance** invention. Don't worry if it's not perfect. The inventor of Monopoly was told that there were 52 things wrong with the game when he first tried to sell it!

Design your board game on a separate piece of paper. When you are happy with the rules, write them here.

UNIT 10: TOPIC 2
Chance experiments

Practice

1 When we look at the chance of something happening, we also need to look at the chance of it **not** happening.

Many people buy lottery tickets. A winner can receive around $30 million! However, the chances of **not** winning the big prize in the lottery are huge.

There is only a one-in-45-million (0.000002%) chance of winning. This means that the chance of **not** winning is 99.999998%!

What are the chances of the following events **not** happening?
Write your answers as fractions.

a Someone chooses "heads" in a coin toss. ________

b Someone rolls a six when they roll a die. ________

2 For this experiment, choose a number from one to nine as your "lucky" number. Write it and three other numbers on a spinner.

a What is the percentage-chance of **not** getting your lucky number? ________

b What is the percentage-chance of getting your lucky number? ________

c If you tried it 40 times, how many times would you expect your lucky number to occur? ________

3 You might spin your lucky number the first time, but there is a greater chance that you will not. Try the experiment with the four numbers from question 2 40 times.

a Write the four numbers in the top row of the table below.

b Start the experiment.

c Fill in the tally chart.

Tally				
Total				

d How did the results for your lucky number compare with your prediction?
If there was a difference, explain why.

____

4 The way an experiment has been set up can alter the results.

Change the spinner so that your lucky number appears **twice** on a spinner with five numbers, and repeat the experiment.

Tally				
Total				

Challenge

1 In a pack of 52 playing cards there are two colours: red and black.

There are four types (called "suits"): hearts, diamonds, clubs and spades.

Use number values to describe the chances of randomly choosing:

a a black card ____________ **b** a diamond ____________

---

2 Based on the number values you wrote in question 1, if somebody took **20** cards at random from the pack, how many of them would you predict to be:

a red? ____________ **b** black? ____________

c diamonds? ____________ **d** hearts? ____________

e clubs? ____________ **f** spades? ____________

---

3 You need a pack of 52 playing cards for this experiment.

- Make sure the cards are shuffled (in a mixed order).
- Place the pile face down.
- Take the top card and look at the type of card. (Red or black? Diamond, heart, club or spade?)
- Write a tally mark in the correct place in the two tables.
- Continue until you have taken 20 cards and recorded the results for all 20 in both tables.
- Fill in the totals.

Type	Red	Black
Tally		
Total		

Type	Diamond	Heart	Club	Spade
Tally				
Total				

---

4 Compare your predictions with the actual results. Explain the difference.

____

---

5 If you were to repeat the experiment, do you think the results would be the same? Give a reason for your answer. (If you have time, repeat the experiment, and check the results!)

____

OXFORD UNIVERSITY PRESS

Mastery

1. We can predict number values, but we can't always say for certain what will happen. For example, if you were to toss a coin six times, it might land on heads every time. For the seventh toss of the coin you might think, "It's got to land on tails this time!" But the chances for the seventh throw would still be 50%—so it might land on tails and it might not! This is called **independent probability**. This means that, no matter how often you repeat an experiment, the chance of something happening remains the same.

   Give another example (or examples) of an experiment in which there is independent probability. ______

---

2. The opposite of independent probability is **dependent probability**. This means that the chance of something happening changes when you repeat the experiment. We can use this bag of marbles as an example.

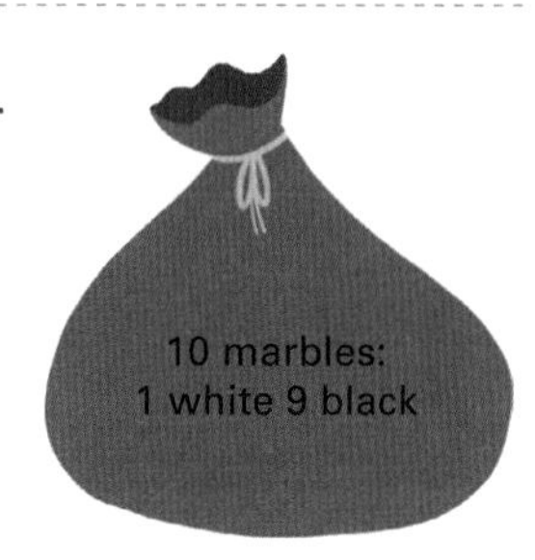

   Write, as a fraction, the chance of:

   a not choosing a white marble ______ **b** choosing a white marble ______

---

3. Eva has the bag of marbles in question 2. She takes out a marble without looking. It is black, but she wants white, so she puts it to one side. The chance of choosing a white marble next time has changed. Explain how. ______

---

4. Eva again picks out a black marble from the nine marbles. What is the chance that she will choose a white marble next time? ______

---

5. If Eva carries on taking black marbles, comment on how the chances will change before there is a 100% chance of picking out the white marble. ______

Dependent probability: an experiment

6. Take your favourite suit from a pack of cards. You should have 13 cards. Think of your favourite card and remember it. Shuffle the cards (or perhaps have someone do it for you) and place them face down.

   a What are the chances of your favourite card being on top? ______

   b What are the chances of your favourite card **not** being on top? ______

---

7. If you were to use the whole pack of 52 cards, how would the chance of your favourite card being on top change?

   ______

ANSWERS

UNIT 1: Topic 1 Place value

Practice

1 a 7000 b 50
c 300 000 d 10 000
e 400 f 2000

2 a forty-seven thousand, six hundred and twenty-five
b one hundred and sixty-two thousand, seven hundred and fifty-two
c three hundred and fourteen thousand, five hundred and twenty
d five hundred and thirteen thousand, eight hundred and four
e five hundred and eighty-two thousand, four hundred
f nine hundred and ninety-two thousand and eight

3 a 30 000 + 2000 + 400 + 80 + 7
b 300 000 + 10 000 + 6000 + 300 + 20 + 1
c 40 000 + 300 + 70
d 800 000 + 6000 + 300 + 2
e 400 000 + 50 000 + 20

Challenge

1 a 980 742 b 897 420
c 24 789 d 98 420.7

2 a 578 472 matches to spike abacus B.
b One million, seven hundred and five thousand, four hundred and seventy-two matches to spike abacus D.
c 857 427 matches to spike abacus A.
d Eight hundred and fifty-seven thousand, two hundred and forty-seven matches to spike abacus C.

3 a 400 000 b 89 000
c 3 740 000 d 1 000 000

Mastery

1

Movie	Amount earned in $	Round to the nearest:	Rounded number in $
The Muppets	165 184 237	Ten thousand	165 180 000
Muppets Most Wanted	80 383 290	Million	80 000 000
The Muppets Movie	65 241 000	Hundred thousand	65 200 000
The Dark Crystal	40 577 001	Hundred thousand	40 600 000
Muppet Treasure Island	34 327 391	Ten thousand	34 330 000
The Great Muppet Caper	31 206 251	Ten thousand	31 210 000
The Muppet Christmas Carol	27 281 507	Thousand	27 282 000
The Muppets Take Manhattan	25 534 703	Million	26 000 000
Muppets from Space	22 323 612	Hundred thousand	22 300 000

2 Students who have an excellent understanding of place value are likely to find many ways of responding to this, such as:
- 100 000 + 50 000 + 7000 + 300 + 10 + 7
- 100 000 + 50 000 + 7000 + 300 + 17
- 100 000 + 50 000 + 7000 + 317
- 100 000 + 50 000 + 7317
- 100 000 + 57 317
- 150 000 + 7000 + 300 + 10 + 7
- 150 000 + 7000 + 300 + 17
- 150 000 + 7000 + 317
- 150 000 + 7317
- 157 000 + 300 + 10 + 7
- 157 000 + 300 + 17
- 157 000 + 317
- 157 300 + 10 + 7
- 157 300 + 17
- 157 310 + 7

UNIT 1: Topic 2 Addition mental strategies

Practice

1 a 279 b 458 c 579
d 2589 e 7789

2 a 145 b 1037 c 1442
d 9505 e 9413 f 10 255

3 a 182 b 339
c 1520 d 3905

4 a 710 b 2550
c 11 100 d 9850

Challenge

1 Teachers to decide whether to ask students to explain their strategies. (Students could share/explain their strategies with each other and with the whole group.)
a 197 b 390 c 447
d 1000 e 2689 f 5900
g 4563 h 6999 i 10 000
j 6954

2 Answers may vary. Possible responses:

Carnivore	Mass (kg)	Rounded mass (kg)	I rounded to the nearest ...
Blue whale	21 364	20 000	Ten thousand
Orca (also known as a killer whale)	9979	10 000	Thousand
Southern elephant seal	4989	5000	Thousand
Walrus	1883	2000	Thousand
Steller sea lion	1102	1100	Hundred
North American brown bear	783	800	Hundred

3 a 12 000 kg (11 862 kg)
b 21 kg
c 2985 kg
d 1343 kg
e A North American brown bear and a Steller sea lion have a combined mass of 1885 kg, 2 kg more than the mass of a walrus.
f 1 kg

Mastery

1 a 60 000
b 60 029
c 111 566

2 This could be carried out as a cooperative group activity and be as simple or as complex as the student/teacher wishes.

UNIT 1: Topic 3 Addition written strategies

Practice

1 a 1919 b 2828 c 3737
d 4646 e 7373 f 8282
g 19 191 h 28 282 i 37 373
j 46 464 k 73 737 l 82 828
m 919 191 n 828 282 o 737 373
p 646 464 q 373 737 r 282 828

Challenge

1 a 43 210 b 54 321 c 65 432
d 654 321 e 765 432

2 a

Country	Area with trees	Area without trees	Total area of the country
Brazil	4 935 380 km²	3 579 497 km²	8 514 877 km²
Canada	3 470 690 km²	6 513 980 km²	9 984 670 km²
USA	3 100 950 km²	6 528 141 km²	9 629 091 km²
China	2 083 210 km²	7 623 751 km²	9 706 961 km²
Australia	1 247 511 km²	6 444 513 km²	7 692 024 km²

b Canada, China, USA, Brazil, Australia

Mastery

1 Teachers might choose to make a calculator available. Students will probably approach the solution through a process of trial and error. However, look for students who see that, since one of the addends is a 5-digit number, the other must be greater than 900 000. For example, 99 999 + 900 001 = 1 000 000. Look also for students who see the connection between addition and subtraction, starting with 1 000 000 and subtracting any 5-digit number (say, 41 449) to find the other addend (in this case 958 551).

2 The answer is 123 456. In order to solve the problem students could use a calculator and start with the next answer (234 567) and subtract two more numbers. For example, 234 567 – 25 478 = 209 089. Subtract 112 741 = 96 348. So, the three addends are 25 478 + 112 741 + 96 348 = 234 567.

3 This could be carried out as a cooperative group activity.

UNIT 1: Topic 4 Subtraction mental strategies

Practice

1 a 55 b 284 c 126
d 347 e 2265 f 3266

2 a 33 b 114 c 231
d 4102 e 4104 f 4322

3 a 425 b 1728
c 217 d 1211

Challenge

1 Students could share their strategies with each other.
a $22.50 b $16.25 c $25.75
d $36.65 e $27.85 f $42.15

2 a 44 b 321 c 46
d 2103 e 2394 f 1250

3 a

River	Place	Length in km	Length in km, rounded to nearest hundred
Amazon	South America	6992	7000
Nile	Africa	6853	6900
Yangtze	China	6380	6400
Mississippi	USA	6275	6300
Yenisei	Russia	5539	5500

b 147 km c 717 km d 100 km

Mastery

1 The wording of answers will vary, but should be similar to the following:
What is the difference between the populations of:
a Nauru and Vatican City?
b Tuvalu and South Ossetia?
c Cook Islands and Palau?
d Monaco and Vatican City?
e Liechtenstein and Palau?

2 Practical activities. These could be carried out as cooperative group activities.

UNIT 1: Topic 5 Subtraction written strategies

Practice

1 a 1221 b 2332 c 3443
d 4554 e 5665 f 6776
g 7887 h 8998 i 12 321
j 23 432 k 34 543 l 45 654
m 56 765 n 67 876 o 123 321
p 234 432 q 345 543 r 456 654

Challenge

1 a 32 123 b 34 234 c 54 345
d 65 456 e 76 567 f 87 678

2 a 164 200 b 1 537 400
c 2 059 500 d 1 395 100
e 16 733 300
f 3 774 900 (14 184 600 – 10 409 700)

Mastery

1 a If the figures from the pages are used, the difference is 1 305 547 000. Students could carry out research to update the figures.
b Students could be asked to check each other's work, perhaps with a calculator.

2 Look for students who see the connection between addition and subtraction as part of their problem-solving strategies.

UNIT 1: Topic 6 Multiplication mental strategies

Practice

1 a 240 b 5430 c 71 230
d 12.5 m e 45 cm f 24 m

2 a 6700 b 2300 c 54 900
d 455 m e 250 cm f $135

3

		× 30	× 40
a	4	4 × 30 = 4 × 3 tens = 12 tens 4 × 30 = 120	4 × 40 = 4 × 4 tens = 16 tens 4 × 40 = 160
b	7	7 × 30 = 7 × 3 tens = 21 tens 7 × 30 = 210	7 × 40 = 7 × 4 tens = 28 tens 7 × 40 = 280
c	9	9 × 30 = 9 × 3 tens = 27 tens 9 × 30 = 270	9 × 40 = 9 × 4 tens = 36 tens 9 × 40 = 360

4

		× 10	Halve it to find × 5	Add the two answers	Multiplication fact
a	24	240	120	240 + 120 = 360	24 × 15 = 360
b	32	320	160	320 + 160 = 480	32 × 15 = 480
c	40	400	200	400 + 200 = 600	40 × 15 = 600
d	50	500	250	500 + 250 = 750	50 × 15 = 750
e	35	350	175	350 + 175 = 525	35 × 15 = 525

Challenge

1 a 200 b 208 c 360
d 512

2 a 200 b 720

3 a 690 b $22

Teacher to check word problems.

Mastery

1 This could be carried out as a cooperative group activity. Alternatively, students could complete their own charts and compare their strategies with others.
Following is an example of how the chart might be completed.

	Strategy	Example	A more difficult example
× 2	Double the number.	16 × 2 = ? 16 + 16 = 32 So, 16 × 2 = 32	42 × 2 = ? 42 + 42 = 84 So, 42 × 2 = 84
× 3	Double the number, then add the answer to the first number.	15 × 3 = ? 15 × 2 = 30 30 + 15 = 45 So, 15 × 3 = 45	35 × 3 = ? 35 × 2 = 70 70 + 35 = 105 So, 35 × 3 = 105
× 4	Double-double (Double the number, then double it again.)	14 × 4 = ? Double: 14 × 2 = 28 Double again: 28 × 2 = 56 So, 14 × 4 = 56	42 × 4 = ? 42 × 2 = 84 Double again: 84 × 2 = 168 So, 42 × 4 = 168
× 5	Multiply by 10, then halve the answer.	18 × 5 = ? 18 × 10 = 180 $\frac{1}{2}$ of 180 = 90 So, 18 × 5 = 90	63 × 5 = ? 63 × 10 = 630 $\frac{1}{2}$ of 630 = 315 So, 63 × 5 = 315
× 6	Multiply by 3, then double the answer.	13 × 6 = ? 13 × 2 = 26 26 + 13 = 39 Double: 39 × 2 = 78 So, 13 × 6 = 78	53 × 6 = ? 53 × 2 = 106 106 + 53 = 159 Double again: 159 × 2 = 318 So, 53 × 6 = 318
× 7	Nobody has thought of a strategy. Can you?	?	?
× 8	Double-double-double (Double it, double it again and then double it again.)	12 × 8 = ? Double: 12 × 2 = 24 Double again: 24 × 2 = 48 Double again: 48 × 2 = 96 So, 12 × 8 = 96	33 × 8 = ? Double: 33 × 2 = 66 Double again: 66 × 2 = 132 Double again: 132 × 2 = 264 So, 33 × 8 = 264
× 9	Multiply the number by 10 and then subtract the first number.	15 × 9 = ? 15 × 10 = 150 150 – 15 = 135 So, 15 × 9 = 135	58 × 9 = ? 58 × 10 = 580 580 – 58 = 522 So, 58 × 9 = 522
× 10	Move everything one place bigger.	24 × 10 = 240	7.85 × 10 = 78.5
× 20, × 30, × 40 etc.	Multiply by the number of tens and then by 10.	16 × 20 = ? 16 × 2 tens = 32 tens 32 × 10 = 320 So, 16 × 20 = 320	16 × 50 = ? 16 × 5 tens = 80 tens 80 × 10 = 800 So, 16 × 50 = 800

ANSWERS

UNIT 1: Topic 7 Multiplication written strategies

Practice

1 a $6 \times 46 = 6 \times 40 + 6 \times 6$
$= 240 + 36$
$= 276$

40 | 6
6 | $6 \times 40 = 240$ | $6 \times 6 = 36$

b $7 \times 37 = 7 \times 30 + 7 \times 7$
$= 210 + 49$
$= 259$

30 | 7
7 | $7 \times 30 = 210$ | $7 \times 7 = 49$

2 a 1685 **b** 2478 **c** 2272 **d** 2572 **e** 879
f 4944 **g** 2142 **h** 4319

Challenge

1 a 211 112 **b** 233 332
c 373 737 **d** 544 445
e 633 336 **f** 936 639

2 a $195.90 **b** 114.8 m
c Students will probably use trial and error. Look for students who estimate by rounding: 6×40 kg = 240 kg; 6×39.5 = 237 kg. Jo can take six bags.
d Students who convert will see that 10 panels would cover 13.20 m, which would be too long.
9×1320 mm = 11 880 mm = 11.88 m.
Audrey needs nine panels.

3 a 345 **b** 756 **c** 882
d 819 **e** 1512 **f** 2241

Mastery

1 a

Insect	Wing beats per second	Wing beats per minute	Wing beats per hour
Midge	1046	62 760	3 765 600
Mosquito	595	35 700	2 142 000
Fruit fly	315	18 900	1 134 000
Wasp	247	14 820	889 200
Fly	190	11 400	684 000

b Answers will vary. Students could check their answers using a calculator.
c 27 216 000 wing beats in a day.

2 a $6 323 400
b This could be as simple or as lengthy as the teacher/student wishes.

UNIT 1: Topic 8 Factors and multiples

Practice

1

a	3	(15)	(21)	23	13	(24)	(12)	28	(33)	43	(30)	(27)	26
b	6	9	(12)	15	(24)	26	32	(30)	16	(36)	20	(42)	(48)
c	4	(8)	14	(16)	22	(24)	18	(32)	30	(40)	(20)	34	(36)
d	9	19	25	(18)	(27)	28	(63)	(36)	60	(54)	49	(90)	39
e	8	28	12	18	(16)	26	(24)	36	(32)	44	(56)	68	(72)

2

a	10	(1)	(2)	3	4	(5)	6	7	8	9	(10)		
b	16	(1)	(2)	3	(4)	5	6	7	(8)	9	10	14	(16)
c	12	(1)	(2)	(3)	(4)	5	(6)	7	8	9	10	11	(12)
d	20	(1)	(2)	3	(4)	(5)	6	7	8	9	(10)	15	(20)
e	27	(1)	2	(3)	4	5	6	7	8	(9)	10	21	(27)
f	18	(1)	(2)	(3)	4	5	(6)	8	(9)	10	12	(18)	20
g	22	(1)	(2)	3	4	5	6	8	9	10	(11)	20	(22)
h	15	(1)	2	(3)	4	(5)	6	8	9	10	12	(15)	20

3 a 1, 2, 4, 8, 16, 32 **b** 1, 11
c 1, 3, 11, 33 **d** 1, 2, 4, 7, 14, 28
e 1, 29 **f** 1, 3, 9
g 1, 2, 6, 7, 21, 42 **h** 1, 7, 49

Challenge

1 a **6**: 1, 2, 3 & 6; **10**: 1, 2, 5 & 10. Common factors are 1 & 2.
b **20**: 1, 2, 4, 5, 10 & 20; **30**: 1, 2, 3, 5, 6, 10, 15 & 30. Common factors are 1, 2, 5 and 10.
c **15**: 1, 3, 5 & 15; **25**: 1, 5, 25. Common factors are 1 and 5.
d **21**: 1, 3, 7 and 21; **27**: 1, 3, 9 & 27. Common factors are 1 and 3.

2 Teacher to check. Answers will vary, for example:
a Because it is an even number.
b Because the sum of the digits is divisible by 6.
c Because it does not end in a zero.
d Because all multiples of 5 end in zero or 5.
e Because it is an even number, and its half is also an even number.
f Because the sum of its digits is not divisible by 6.

3 a 30 **b** 12 **c** 36
d 24 **e** 90 **f** 77

4 36

5 72

Mastery

1 a 1, 12, 2, 6, 3, 4
b 16

2 a 18 is the abundant number.
$(1 + 2 + 3 + 6 + 9 = 21)$
16 is not an abundant number.
$(1 + 2 + 4 + 8 = 15)$
b This could be carried out as a cooperative group activity. The abundant numbers to 100 are:
12, 18, 20, 24, 30, 36, 40, 42, 48, 54, 56, 60, 66, 70, 72, 78, 80, 84, 88, 90, 96, 100.

UNIT 1: Topic 9 Divisibility

Practice

1 a Student circles 45, 70, 310, 990, 2995, 4440, 7110.
b Student circles 18, 72, 810, 4314, 3054, 2226.

2 a Teacher to check.
b Student circles 8, 32, 72, 168, 880, 2168, 1248, 3880, 7832, 3488, 5728.

3 Teacher to check. Look for students who follow the rules for divisibility on page 41 of the student book.

4 Teacher to check. Answers will vary, e.g. "Because the sum of the digits is divisible by 3".

Challenge

1 a Teacher to check. Answers will vary, e.g. "Because it is only divisible by 1 and itself".
b 47

2 a 51, 54, 57, 60 **b** 55, 60
c 56 **d** 54
e 54, 60 **f** 52, 56, 60
g 54 **h** 56

3 a 72
   b 1, 72, 2, 36, 3, 24, 4, 18, 6, 12, 8, 9
   c 71, 73 and 79

4 This could be carried out as a cooperative group activity.
   a The numbers are 60, 72, 84, 90 and 96. The factors are:
      - 60: 1, 60, 2, 30, 3, 20, 4, 15, 5, 12, 6, 10
      - 72: 1, 72, 2, 36, 3, 24, 4, 18, 6, 12, 8, 9
      - 84: 1, 84, 2, 42, 3, 28, 4, 21, 6, 14, 7, 12
      - 90: 1, 90, 2, 45, 3, 30, 5, 18, 6, 15, 9, 10
      - 96: 1, 96, 2, 48, 3, 32, 4, 24, 6, 16, 8, 12
   b 1, 2,3 and 6

Mastery

1 a **3**: 3, 6, 9, 12
     4: 4, 8, 12
     The lowest common multiple of 3 and 4 is 12.
   b **3**: 3, 6, 9, 12, 15, 18, 21, 24
     8: 8, 16, 24
     The lowest common multiple of 3 and 8 is 24.

2 a **3**: 3, 6, 9, 12, 15, 18, 21, 24
     4: 4, 8, 12, 16, 20, 24
     8: 8, 16, 24
     The lowest common multiple of 3, 4 and 8 is 24.
   b **6**: 6, 12, 18, 24, 30, 36, 42, 48, 54, 60
     15: 15, 30, 45, 60
     20: 20, 40, 60
     The lowest common multiple of 6, 15 and 20 is 60.

3 120

UNIT 1: Topic 10 Division written strategies

Practice

1 a 142 ÷ 2 is the same as
     100 ÷ 2 and 42 ÷ 2
     100 ÷ 2 = 50
     42 ÷ 2 = 21
     So, 142 ÷ 2 = 50 + 21 = 71
   b 128 ÷ 4 is the same as
     100 ÷ 4 and 28 ÷ 4
     100 ÷ 4 = 25
     28 ÷ 4 = 7
     So, 128 ÷ 4 = 25 + 7 = 32
   c 135 ÷ 5 is the same as
     100 ÷ 5 and 35 ÷ 5
     100 ÷ 5 = 20
     35 ÷ 5 = 7
     So, 135 ÷ 5 = 20 + 7 = 27
   d 324 ÷ 3 is the same as
     300 ÷ 3 and 24 ÷ 3
     300 ÷ 3 = 100
     24 ÷ 3 = 8
     So, 324 ÷ 3 = 100 + 8 = 108

2 a 43 b 22 c 32
   d 34 e 216 f 151
   g 121 h 122 i 1634
   j 1146 k 1053 l 1392
   m 12 696 n 11 541 o 14 507

Challenge

1 a 7 b 8 c 9
   d 9 e 34 f 64
   g 39 h 99 i 393
   j 819 k 847 l 654

2 a 12 r3 b 15 r3 c 10 r2
   d 13 r4 e 147 r2 f 308 r1
   g 119 r3 h 102 r5 i 1336 r2
   j 714 r4 k 1391 r2 l 1873 r3
   m 28 298 r2 n 3238 r3 o 6148 r7

Mastery

1 Teachers may wish to ask students to explain the way they have used/not used the remainder.
   a 3875 marbles. (The remainder of 5 cannot be shared.)
   b 3876 cartons. (The last carton will contain only 5 cups.)
   c $3875. (Students could use decimals, in which case $3875.80 is the amount they will each get, and 20 cents will be left over.)

2 **a** and **b**

Airport	Number of passengers per day, to the nearest whole number
Hartsfield–Jackson, USA	278 819 (rounded up)
Beijing, China	247 088 (rounded up)
Dubai, UAE	214 313 (rounded down)
Chicago, USA	211 374 (rounded up)
Tokyo, Japan	206 923
London Heathrow, UK	206 016 (rounded down)
Paris CDG, France	180 687 (rounded up)
Sydney, Australia	108 945

UNIT 2: Topic 1 Comparing and ordering fractions

Practice

1 Student shades:
   a one section b five sections
   c three sections d three sections
   e three sections

2 Student shades:
   a seven shapes b six shapes
   c four shapes d two shapes

3 a $\frac{3}{6}$ or $\frac{1}{2}$ b $\frac{4}{8}$ or $\frac{1}{2}$
   c $\frac{3}{10}$ d $\frac{7}{12}$

4 Student shades appropriate number of sections and writes that:
   a $\frac{1}{3}$ is bigger. b $\frac{1}{5}$ is bigger.
   c $\frac{1}{6}$ is bigger. d $\frac{1}{9}$ is bigger.

Challenge

1

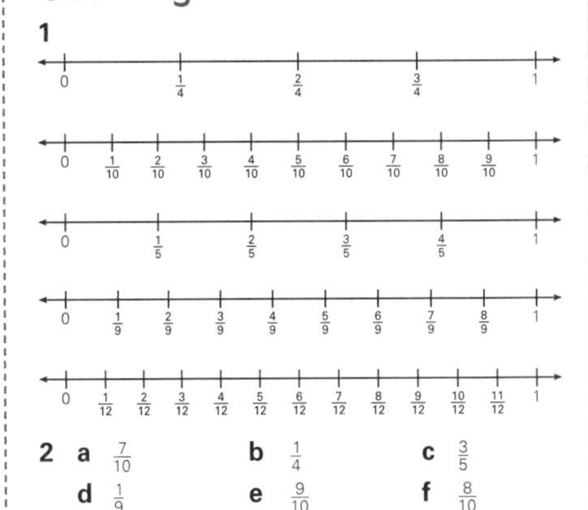

2 a $\frac{7}{10}$ b $\frac{1}{4}$ c $\frac{3}{5}$
   d $\frac{1}{9}$ e $\frac{9}{10}$ f $\frac{8}{10}$

3 a $\frac{1}{10}, \frac{1}{5}, \frac{3}{10}, \frac{2}{5}, \frac{1}{2}, \frac{6}{10}, \frac{10}{10}$
   b $\frac{1}{12}, \frac{1}{10}, \frac{1}{4}, \frac{4}{12}, \frac{4}{10}, \frac{4}{5}, 1$
   c $\frac{1}{12}, \frac{1}{10}, \frac{1}{9}, \frac{1}{5}, \frac{1}{4}, \frac{1}{2}, 1$

Mastery

1 Practical activities

2 Multiple answers possible. For example, Sean may have bought $4\frac{1}{3}$ cakes and be giving 13 people each one slice. Look for students who show an understanding of mixed numbers and how they relate to individual fractional parts.

3 Multiple answers possible. Look for students who can correctly interpret the problem and who are able to accurately convert between improper fractions and mixed numbers.

UNIT 2: Topic 2 Adding and subtracting fractions

Practice

1 a $\frac{5}{8} + \frac{2}{8} = \frac{7}{8}$ b $\frac{3}{6} + \frac{2}{6} = \frac{5}{6}$
   c $\frac{3}{4} - \frac{1}{4} = \frac{2}{4}$ d $\frac{3}{5} - \frac{1}{5} = \frac{2}{5}$

2 Accept equivalent fractions. Teacher to check shading.
   a $\frac{6}{8} + \frac{1}{8} = \frac{7}{8}$ b $\frac{5}{10} - \frac{2}{10} = \frac{3}{10}$
   c $\frac{1}{6} + \frac{2}{6} = \frac{3}{6}$ d $\frac{1}{3} + \frac{1}{3} = \frac{2}{3}$
   e $\frac{4}{4}$ (or 1) $- \frac{2}{4}$ (or $\frac{1}{2}$) $= \frac{2}{4}$ (or $\frac{1}{2}$)
   f $\frac{4}{6} - \frac{1}{6} = \frac{3}{6}$ (or $\frac{1}{2}$)

3 Teacher to check shading.
   a $\frac{1}{5} + \frac{2}{5} = \frac{3}{5}$ b $\frac{3}{6} + \frac{2}{6} = \frac{5}{6}$
   c $\frac{2}{9} + \frac{5}{9} = \frac{7}{9}$ d $\frac{1}{8} + \frac{3}{8} = \frac{4}{8}$ (or $\frac{1}{2}$)

4 Some students may choose to complete the questions by crossing out sections of the shapes instead of shading them.
   a $\frac{7}{10} - \frac{3}{10} = \frac{4}{10}$ b $\frac{6}{7} - \frac{3}{7} = \frac{3}{7}$
   c $\frac{4}{5} - \frac{1}{5} = \frac{3}{5}$ d $\frac{2}{3} - \frac{2}{3} = 0$

Challenge

1 Teacher to check shading. Accept equivalent fractions.
   a $\frac{2}{3} + \frac{2}{3} = \frac{4}{3} = 1\frac{1}{3}$
   b $\frac{5}{6} + \frac{5}{6} = \frac{10}{6} = 1\frac{4}{6}$
   c $\frac{3}{4} + \frac{3}{4} = \frac{6}{4} = 1\frac{2}{4}$ (or $1\frac{1}{2}$)
   d $\frac{2}{8} + \frac{5}{8} + \frac{4}{8} = \frac{11}{8} = 1\frac{3}{8}$

ANSWERS

2 **a** $\frac{4}{4}$ (or 1) **b** $\frac{5}{8}$
c $\frac{14}{10} = 1\frac{4}{10}$ **d** $\frac{9}{6} = 1\frac{3}{6}$ (or $1\frac{1}{2}$)
e $\frac{11}{5} = 2\frac{1}{5}$ **f** $1\frac{1}{4}$
g $\frac{32}{10} = 3\frac{2}{10}$

Mastery

1 Answers will vary. Students could share the answers with each other. Alternatively, a group could be given the task of finding as many solutions as possible.

2 Answers may vary. For example:
a $\frac{7}{8} + \frac{7}{8} > 1$ **b** $1\frac{1}{10} + \frac{1}{10} < 1\frac{3}{10}$
c $\frac{1}{4} + \frac{1}{2} < 1$ **d** $\frac{1}{10} + \frac{1}{10} = \frac{1}{5}$

3 Answers may vary. For example:
$\frac{4}{4} - \frac{1}{4} = \frac{3}{4}$
$\frac{5}{4} - \frac{1}{2} = \frac{3}{4}$

4 This could be given as a cooperative group task. The total must equal two pizzas. For example: Tran eats $\frac{1}{4}$, Jim eats $\frac{1}{2}$, Sam eats $1\frac{1}{4}$.

5 Teachers may wish to set this as a group challenge and to make circles available on spare paper. Successful responses could be shared with the whole group and/or displayed.

UNIT 2: Topic 3 Decimal fractions

Practice

1 **a** $\frac{7}{10}$ (or $\frac{70}{100}$), 0.7
b $\frac{7}{100}$, 0.07
c $\frac{53}{100}$, 0.53

2 **a** Student shades 20 small squares and writes $\frac{2}{10}$ or $\frac{20}{100}$.
b Student shades 17 small squares and writes $\frac{17}{100}$.
c Student shades 1 small square and writes $\frac{1}{100}$.
d Student shades 69 small squares and writes $\frac{69}{100}$.

3 **a** 0.007, $\frac{7}{1000}$ **b** 0.013, $\frac{13}{1000}$

4 **a** 0.115 **b** 0.87 **c** 0.017

5 **a** $\frac{3}{1000}$ **b** $\frac{45}{100}$ **c** $\frac{862}{1000}$

Challenge

1 **a** $0.001 < 0.01$ **b** $\frac{5}{100} > 0.005$
c $\frac{350}{1000} = 0.35$ **d** $0.01 > 0.009$
e $\frac{25}{1000} < 0.25$ **f** $\frac{10}{1000} = 0.01$
g $0.99 > \frac{99}{1000}$ **h** $1\frac{5}{10} > 1.05$
i $\frac{199}{1000} < 0.2$

2 **a** 0.1, 0.2, 0.3, 0.4, 0.5, 0.6, 0.7, 0.8, 0.9
b 3.1, 3.2, 3.3, 3.4, 3.5, 3.6, 3.7, 3.8, 3.9
c 0.11, 0.12, 0.13, 0.14, 0.15, 0.16, 0.17, 0.18, 0.19
d 0.031, 0.032, 0.033, 0.034, 0.035, 0.036, 0.037, 0.038, 0.039

3 **a** 0.3, 0.4, $\frac{7}{10}$, 0.8, $\frac{9}{10}$
b 0.01, $\frac{2}{100}$, 0.05, 0.07, $\frac{3}{10}$
c 0.008, $\frac{10}{1000}$, 0.08, 0.1, 0.8
d $\frac{5}{1000}$, $\frac{15}{1000}$, 0.05, 0.051, 0.5
e $\frac{4}{100}$, 0.4, $\frac{404}{1000}$, 0.444, 4.004
f 0.025, 0.25, 0.255, 2.5, 2.55

Mastery

1 This would make an ideal cooperative group task. Teachers who would like students to make decimal fractions to more than 3 places may wish to provide students with large squares of paper (perhaps 1 m^2).

2 The first six places are:
$\frac{4}{10}$
$\frac{2}{100}$
$\frac{8}{1000}$
$\frac{5}{10\,000}$
$\frac{7}{100\,000}$
$\frac{1}{1\,000\,000}$
Students who have a good understanding of place value should be able to identify the value of each decimal digit.

UNIT 2: Topic 4 Percentages

Practice

1 **a** $\frac{99}{100}$, 0.99, 99%
b $\frac{9}{100}$, 0.09, 9%
c $\frac{9}{10}$ ($\frac{90}{100}$), 0.9, 90%

2 Student shades:
a 45 squares and writes $\frac{45}{100}$ and 45%.
b 40 squares and writes 0.4 and 40%.
c 34 squares and writes $\frac{34}{100}$ and 34%.
d 5 squares and writes 0.05 and 5%.

3 Teachers may wish to discuss equivalent fractions with students (e.g. $\frac{10}{100}$ instead of 10%).

	Fraction	Decimal	Percentage
a	$\frac{15}{100}$	0.15	15%
b	$\frac{1}{10}$	0.1	10%
c	$\frac{85}{100}$	0.85	85%
d	$\frac{5}{100}$	0.05	5%
e	$\frac{39}{100}$	0.39	39%
f	$\frac{7}{100}$	0.07	7%

4 **a** 35% = 0.35 **b** 0.19 < 20%
c $\frac{5}{10} > 0.05$ **d** 90% = $\frac{9}{10}$
e 0.7 > 68%

Challenge

1 **a** 4%, 0.045, $\frac{14}{100}$, 0.4
b 0.72, 73%, 74%, $\frac{3}{4}$
c 0.03, $\frac{13}{100}$, $\frac{3}{10}$, 33%
d 0.095, 95%, 9.05, 9.5
e 0.015, 5%, $\frac{15}{100}$, 1.5
f 0.024, 2.5%, $\frac{23}{100}$, $\frac{1}{4}$

2 Look for students who see that, since 10% of 40 is 4 stars, 20% = 2 × 4 = 8 stars.
a Student shades 4 stars red, 10 stars blue, 10 stars green and 8 stars yellow.
b There are 8 stars left, which is $\frac{1}{5}$ (accept $\frac{8}{40}$, $\frac{4}{20}$, $\frac{2}{10}$), 0.2 and 20%.

3 **a** The simplest solution is to start with four simple percentages that will total 100%. For example, red = 20%, blue = 60%, yellow and green = 10% each. Total is 100%.
10% of 50 is 5 stars. So, 5 stars are green and 5 stars are yellow. 20% are red, which is double 10%. 5 stars × 2 = 10 stars. From this the student can either work out that 60% = 6 × 5 stars = 30 stars or simply colour the remaining 30 stars blue.
b
- blue: 60% = 30 stars
- red: 20% = 10 stars
- green: 10% = 5 stars
- yellow: 10% = 5 stars

Mastery

1 Answers may vary. Students could share their ideas with each other. Example response: 100% means the whole amount. It is impossible to give more than 100% of your energy or effort.

2 **a** 10% off
b 25% discount
c Save 15%
d Contains 20% less sugar
e Up to 70% off
f Buy two and save 5%

3 **a & b** Answers may vary. Students could carry this out as a think-pair-share activity.
c Practical activity

UNIT 3: Topic 1 Financial plans

Practice

1 $20 per student

2 **a** $1.20 **b** $9.95
c $23.95 **d** $12.45
e $30 **f** $9.75

3 **a** $3
b Answers will vary but should total $15.

4 **a** $2.95 **b** $3.80

Challenge

1 **a** Discount: $0.80. New price: $7.20 (accept $7.19)
b Discount: $1.50. New price: $13.50 (accept $13.49)
c Discount: $0.50. New price: $1.50 (accept $1.49)
d Discount: $0.30. New price: $0.30 (accept $0.29)
e Discount: $0.25. New price: $2.25 (accept $2.24)
f Discount: $3.60 (off $36). New price: $32.40 for 20

2 **a** 25 bats = $199.75
25 ropes = $49.75
25 hoops = $62.25
25 packs of balls = $374.75
25 × 4 small balls = $59
Total = $745.50
b This is $245.50 more than they have.

3 **a** The new total is $460.70.
b They would have $39.30 left from their budget of $500.

Mastery

Answers will vary. This would make an ideal cooperative group activity.

UNIT 4: Topic 1 Number patterns

Practice

1 a

Term	1	2	3	4	5	6	7	8	9	10
Number	0	1.5	3	4.5	6	7.5	9	10.5	12	13.5

b

Term	1	2	3	4	5	6	7	8	9	10
Number	5	$4\frac{3}{4}$	$4\frac{1}{2}$	$4\frac{1}{4}$	4	$3\frac{3}{4}$	$3\frac{1}{2}$	$3\frac{1}{4}$	3	$2\frac{3}{4}$

2 Answers for the rules may vary. Examples include:

a 30, 27.5, 25, 22.5, 20, 17.5, 15, 12.5, 10, 7.5
Rule: *Start at 30. Decrease by 2.5 each time.*

b 10, 9.75, 9.5, 9.25, 9, 8.75, 8.5, 8.25, 8, 7.75
Rule: *Start at 10. Decrease by 0.25 each time.*

c 0, $\frac{1}{3}$, $\frac{2}{3}$, 1, $1\frac{1}{3}$, $1\frac{2}{3}$, 2, $2\frac{1}{3}$, $2\frac{2}{3}$, 3
Rule: *Start at 0. Increase by $\frac{1}{3}$ each time.*

d 10, 9.6, 9.2, 8.8, 8.4, 8, 7.6, 7.2, 6.8, 6.4
Rule: *Start at 10. Decrease by 0.4 each time.*

3 Rules may vary. Alternative rule for question c could be:

Start with 1 stick. Add 4 sticks for each pentagon.

Pattern of sticks	Rule for making the pattern	How many sticks are needed?					
a	Start with 5 sticks. Increase the number of sticks by 5 for each new pentagon.	Number of pentagons	1	2	3	4	5
		Number of sticks	5	10	15	20	25
b	Start with 4 sticks. Increase the number of sticks by 3 for each new diamond.	Number of diamonds	1	2	3	4	5
		Number of sticks	4	7	10	13	16
c	Start with 5 sticks. Increase the number of sticks by 4 for each new pentagon.	Number of pentagons	1	2	3	4	5
		Number of sticks	5	9	13	17	21

Challenge

1 Wording of students' answers may differ, but steps should lead to 1.

a The digit sum of 35 is 3 + 5 = 8.
8 is not divisible by 3, so 35 – 1 = 34.
The digit sum of 34 is 7.
7 is not divisible by 3, so 34 – 1 = 33.
The digit sum of 33 is 6.
6 is divisible by 3, so 33 ÷ 3 = 11.
The digit sum of 11 is 2.
2 is not divisible by 3, so 11 – 1 = 10.
The digit sum of 10 is 1.
1 is not divisible by 3, so 10 – 1 = 9.
9 is divisible by 3, so 9 ÷ 3 = 3.
The digit sum of 3 is 3.
3 ÷ 3 = 1
STOP.

b The digit sum of 362 is 3 + 6 + 2 = 11.
11 is not divisible by 3, so 362 – 1 = 361
The digit sum of 361 is 10.
10 is not divisible by 3, so 361 – 1 = 360.
The digit sum of 360 is 9.
9 is divisible by 3, so 360 ÷ 3 = 120.
The digit sum of 120 is 3.
3 is divisible by 3, so 120 ÷ 3 = 40.
The digit sum of 40 is 4.
4 is not divisible by 3, so 40 – 1 = 39.
The digit sum of 39 is 12.
12 is divisible by 3, so 12 ÷ 3 = 4.
4 is not divisible by 3, so 4 – 1 = 3.
The digit sum of 3 is 3.
3 ÷ 3 = 1
STOP.

2 Students could share their findings with each other.

Mastery

1 Students could work on their drafts as part of cooperative group work. This is an example of a finished flow chart:

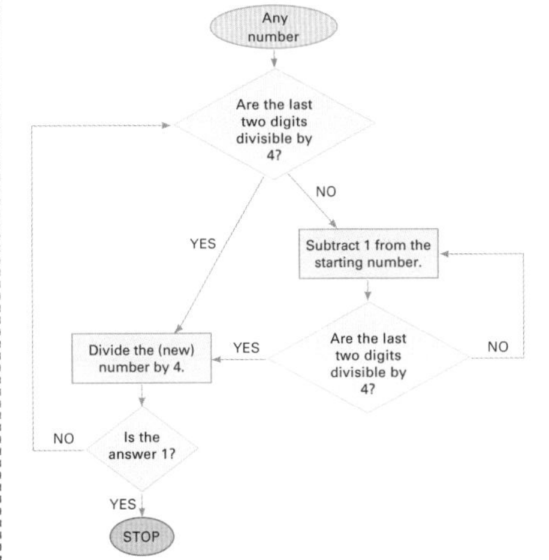

UNIT 4: Topic 2 Number operations and properties

Practice

1 a Addition: Changing the order of the numbers **does** give the same answer.

b Subtraction: Changing the order of the numbers **does not** give the same answer.

c Multiplication: Changing the order of the numbers **does** give the same answer.

d Division: Changing the order of the numbers **does not** give the same answer.

2 Students could use mental strategies that do not necessitate changing the order of the numbers. Teachers could ask students to explain their method of solving the problem.

a 25 + 15 = 40 + 18 = 58
b 20 × 5 = 100 × 16 = 1600
c 47 + 13 = 60 + 37 = 97
d 500 × 2 = 1000 × 17 = 17 000

3 Teachers could ask students to explain their method of solving the problems. Possible responses:

a 43 – 23 = 20, 20 – 16 = 4
b 81 ÷ 9 = 9, 9 ÷ 3 = 3
c 69 – 19 = 50, 50 – 25 = 25
d 96 ÷ 8 = 12, 12 ÷ 3 = 4

Changing the order of the two numbers being subtracted or divided **does** give the same answer.

Challenge

1 a 25 × 4 = 65 + 35
b 125 ÷ 5 = 75 ÷ 3
c 48 ÷ 3 = 8 × 2
d 90 – 30 = 15 × 4
e 100 ÷ 5 = 19 + 1
f 65 + 35 = 200 ÷ 2
g 25 × 6 = 15 × 10
h 110 – 20 = 15 × 6
i 120 ÷ 3 = 80 ÷ 2

2 Multiple answers are possible. Students could offer their solutions to others to check whether they balance. Teachers may wish to encourage students to use all four operations in solving the problems.

Mastery

1 a and **b** Teacher to check explanation. Look for students who recognise that changing the order of the numbers makes the multiplication easier because one of the multiplicands is 10 (i.e. 5 × 2 = 10, 9 × 4 = 36, 10 × 36 = 360).

2 The mystery number is 6.

a 24 ÷ 6 = 24 ÷ 2 – 8
b 24 × 2 = 8 × 6
c (24 – 6) + 24 ÷ 6 = 8 + 8 + 8 – 2
d (24 ÷ 8) × 2 – 6 = 24 – 8 – 8 – 2^2 × 2

3 a and **b** Answers will vary. Students could try their puzzles out on each other.

Another example of using 6 as the mystery number and the numbers 24, 2 and 8 is:
24 ÷ ◆ + 8 = 24 ÷ 2, which becomes
24 ÷ 6 + 8 = 24 ÷ 2.

UNIT 5: Topic 1 Length and perimeter

Practice

1 Teacher to decide on level of tolerance. Suggest +/– 1 mm.

a 7 cm 3 mm or 7.3 cm
b 7 cm 9 mm or 7.9 cm
c 14 cm 5 mm or 14.5 cm
d 11 cm 2 mm or 11.2 cm
e 13 cm 6 mm or 13.6 cm
f 8 cm 8 mm or 8.8 cm
g 10 cm 1 mm or 10.1 cm

2 a 22 cm **b** 16 cm **c** 18 cm

ANSWERS

3

	Centimetres	Millimetres
a	5 cm	50 mm
b	5.5 cm	55 mm
c	27 cm	270 mm
d	3.8 cm	38 mm
e	2.2 cm	22 mm

4

	Metres	Centimetres
a	3 m	300 cm
b	2.5 m	250 cm
c	7.35 m	735 cm
d	1.25 m	125 cm
e	$3\frac{1}{2}$ m	350 cm

5

	Kilometres	Metres
a	5 km	5000 m
b	3.5 km	3500 m
c	7.25 km	7250 m
d	4.75 km	4750 m
e	1.1 km	1100 m

Challenge

1 Teacher to decide and check on the level of accuracy. The two units of length should be appropriate to the size of the object being measured.

2 Teacher to decide on acceptable level of accuracy. Suggest +/– 1 mm per side.
  a 4×2.2 cm = 88 mm or 8.8 cm
  b 5×2.7 cm = 135 mm or 13.5 cm
  c 2×2.8 cm + 1 cm = 66 mm or 6.6 cm
  d 3 cm + 2.5 cm + 1.2 cm + 1.7 cm = 83 mm or 8.3 cm

3 Students will need to recognise that each side should be 4.5 cm long. They should find the base line simple to draw but will probably need more than one attempt to get the second and third sides correct. Those who are familiar with using a protractor may decide to use one.

4 Students should recognise that a line does not need to be straight, nor does it need to continue in the same direction along its length.

Mastery

1

Type of eagle	Wingspan (mm)	Wingspan (cm and mm)	Wingspan (cm)	Wingspan (m)
Wedge-tailed eagle	2840 mm	284 cm 0 mm	284 cm	2.84 m
Golden eagle	2805 mm	280 cm 5 mm	280.5 cm	2.805 m
Martial eagle	2625 mm	262 cm 5 mm	262.5 cm	2.625 m
Sea eagle	2490 mm	249 cm 0 mm	249 cm	2.49 m
Bald eagle	2275 mm	227 cm 5 mm	227.5 cm	2.275 m

2 Answers may vary. Students could justify their choice of unit to each other. Possible responses:
  a cm or m b cm or mm c mm

3 This could be carried out as a cooperative group assignment and could be as detailed or as simple as the student/teacher desires.

UNIT 5: Topic 2 Area

Practice

1 a $18\ cm^2$ b $32\ cm^2$ c $16\ cm^2$
  d $35\ cm^2$ e $10\ cm^2$ f $12\ cm^2$
  g $21\ cm^2$ h $16\ cm^2$

Challenge

1 a $15\ m^2$ b $12\ m^2$
  c $8\ m^2$ d $35\ m^2$

2 Look for students who understand the need to look for rectangles in their calculations. Some may choose to find the whole area and subtract the "missing" rectangle(s). Others may opt for splitting the shapes into an appropriate number of rectangles. Students could share their strategies with each other.
  a $19\ cm^2$ b $23\ cm^2$

3 a $26\ m^2$ b $20\ m^2$ c $6\ m^2$

Mastery

1

State/territory	Area (rounded)
WA	2.5 million km^2
QLD	1.7 million km^2
NT	1.3 million km^2
SA	980 000 km^2
NSW	800 000 km^2
VIC	225 000 km^2
TAS	68 000 km^2
ACT	2000 km^2

  b The sum of the areas as shown is 7 575 000 km^2.

2 Students may choose to calculate the area of the triangle by counting full and half squares. Others may see the triangle as having half the area of an imaginary rectangle drawn around it. The area of the triangle shown is $\frac{1}{2}$ of $12\ cm^2 = 6\ cm^2$.

3 Practical activity. Students will probably need a calculator to work out the exact area of the room they choose.

UNIT 5: Topic 3 Volume and capacity

Practice

1 a 120 L b 200 mL
  c 45 L d 5 mL

2 a $12\ cm^3$ b $24\ cm^3$ c $16\ cm^3$

3 a Teacher to check. Students could explain that 18 1-centimetre cubes could be used to make a model with the same dimensions.
  b $72\ cm^3$

4 a $48\ cm^3$ b $72\ cm^3$

Challenge

1 a Students could share their responses with each other. Likely responses are: 1 L 750 mL, 1750 mL, 1.75 L
  b 5.25 L = 5250 mL

2 a $90\ cm^3$ b $180\ cm^3$ c $500\ cm^3$
  d $600\ cm^3$ e $240\ cm^3$ f $324\ cm^3$

3 a 250 mL
  b 1000 mL or 1 L
  c 3750 mL or 3.75 L

Mastery

1 Practical activities. The following information about approximate quantities of daily water usage could be useful:
- Toilet flush: 25 L
- Washing hands: 5 L
- Brushing teeth: 5–10 L
- Bath: up to 150 L
- Handwashing dishes: 10 L
- Dishwasher: 20 L
- Washing machine: 100 L
- Cooking: 1 L
- Water for drinking: 2 L (average three cups per day)
- Dripping tap: 5 L

2 This could be carried out as a cooperative group assignment and could be as detailed or as simple as the student/teacher desires.

UNIT 5: Topic 4 Mass

Practice

1

	Tonnes	Kilograms
a	3 t	3000 kg
b	5 t	5000 kg
c	2.5 t	2500 kg
d	4.25(0) t	4250 kg
e	2.75 t	2750 kg

	Kilograms	Grams
a	7 kg	7000 g
b	4 kg	4000 g
c	1.25(0) kg	1250 g
d	5.75 kg	5750 g
e	0.25 kg	250 g

	Grams	Milligrams
a	1 g	1000 mg
b	2 g	2000 mg
c	7.5 g	7500 mg
d	1.25 g	1250 mg
e	0.75 g	750 mg

2 Answers will vary but need to be appropriate to the unit of mass. For example:
 a a pinch of salt b a banana
 c a person d a truck

3 a 200 g b 700 g
 c 1100 g d 1700 g

4 a 0 kg 500 g, 0.5 kg
 b 1 kg 600 g, 1.6 kg
 c 2 kg 750 g, 2.75 kg
 d 0 kg 750 g, 0.75 kg

Challenge

1 Teacher to decide on level of accuracy required. Pointer for 1b to be approximately halfway between 1.2 kg and 1.3 kg. Pointer for 1d to be approximately halfway between 2.5 kg and 3 kg.

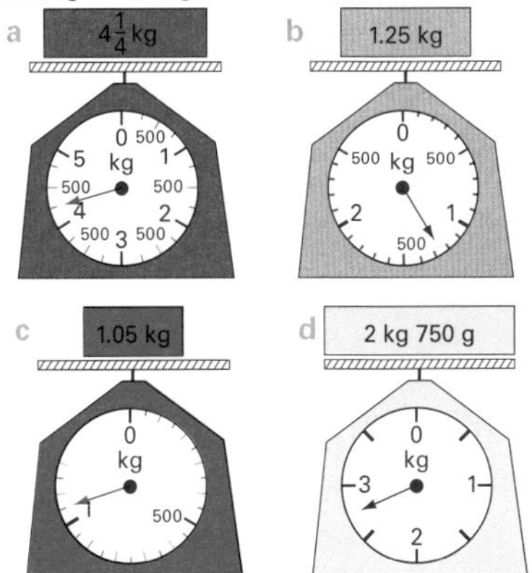

2 Teacher to check. Students could share their responses with each other. Likely responses:
 a $4\frac{1}{4}$ kg, 4.25 kg, 4 kg 250 g, 4250 g
 b $1\frac{1}{4}$ kg, 1.25 kg, 1 kg 250 g, 1250 g
 c $1\frac{5}{100}$ kg or $1\frac{50}{1000}$ kg, 1.05 kg, 1 kg 50 g, 1050 g
 d $2\frac{3}{4}$ kg, 2.75 kg, 2 kg 750 g, 2750 g

3 a Billy, Jo, Mo, Eva, Alex, Dee
 b 27.85 kg
 c 174 kg ÷ 6 = 29 kg
 d Eva
 e 1100 g
 f Dee and Alex (59.55 kg)

4 Teacher to check. Look for students who write appropriate masses, e.g.

Item	Mass	Item	Mass
Small bottle of water	350 g	Carrot sticks	100 g
Large bottle of water	1 kg	Apple	200 g
Bottle of juice	250 g	Bunch of grapes	100 g

Mastery

1 190 tonnes = 190 000 kg. Subtract 600 g = 189 999 kg 400 g or 189 999.4 kg

2 a Bowhead whale, North Atlantic right whale, North Pacific right whale, Southern right whale, grey whale, humpback whale
 b 9144 kg or 9 tonnes 144 kg or 9.144 tonnes
 c The Bowhead and the North Pacific right whales have a combined mass of 189.202 tonnes.
 d 189 999.4 kg – 189 202 kg = 797.4 kg

3 This could be carried out as a cooperative group activity.

UNIT 5: Topic 5 Time

Practice

1

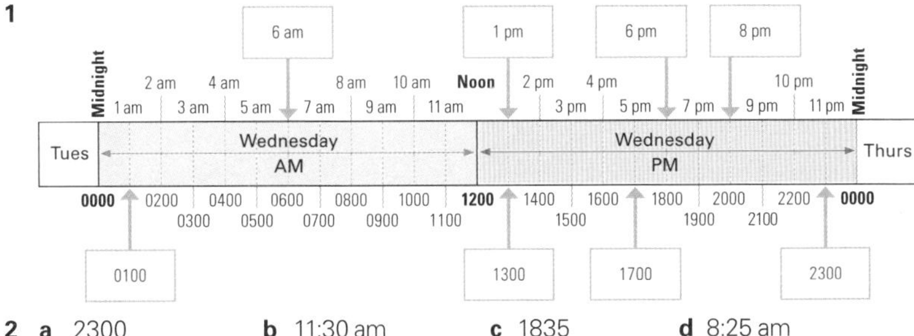

2 a 2300 b 11:30 am c 1835 d 8:25 am
 e 9:40 pm f 1922 g 0444 h 12:05 am

3 Answers will vary but should be appropriate to the relevant activity.

4 a

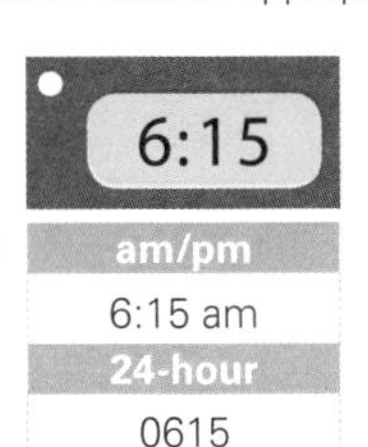

b

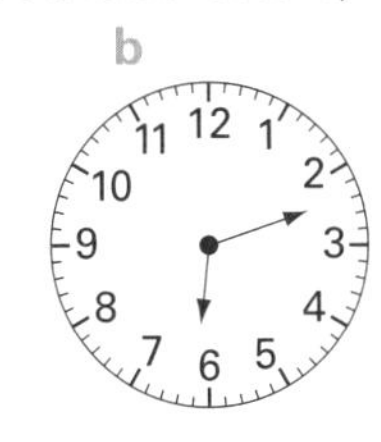

6:12
am/pm
6:12 pm
24-hour
1812

Challenge

1 a

 b 20 minutes c 1815
 d 50 minutes e 1430 to 1440
 f 1200 g 11:55 am
 h Five and a half hours
 i Answers will vary but are likely to be in the "Quiet time" session (between 1430 and 1440).
 j 3 hours 35 minutes
 k Answers will vary, but should total 3 hours and 35 minutes.

Mastery

1 a

Town	Time of arrival/ departure
Tinytown	2200
Smallville	0110
Bigtown	0140
Medium City	0400
Largetown	0515
Great City	0540
Hugeville	0610

 b Tinytown and Smallville
 c Largetown to Great City (25 minutes)
 d 8 hours and 10 minutes

2 a 10 am b 11 pm c 3 am
 d 5 am e 7.30 pm f 6 pm

3 Teacher to check. Look for students who understand the concept of time zones and who can accurately calculate world times in relation to Greenwich Mean Time.

UNIT 6: Topic 1 2D shapes

Practice

1 a A: quadrilateral; B: octagon; D: pentagon; F: hexagon; H: triangle
 b Teacher to check explanation. Likely responses include:
  - "Drawing C is not a polygon because it is not a closed shape."
  - "Drawing E is not a polygon because all the lines are not straight."
  - "Drawing G is not a polygon because the lines cross each other."

2 Student shades shapes b, d, e and j. (Shape g is not a polygon.)

3 a equilateral b scalene c right-angled

ANSWERS

Challenge

1 Teacher to decide on level of accuracy required. Note that students could choose to draw a rectangle for an irregular quadrilateral or a parallelogram. A square is also a rectangle and a parallelogram.

2 Teacher to check. Likely responses include:
   - **a** "I know it is a square because all the angles and all the sides are equal."
   - **b** "I know it is a trapezium because there is a pair of parallel sides."
   - **c** "I know it is a rhombus because it has four equal sides and the opposite angles are equal."

3 Note: shape c could also be a rhombus. Likely responses are:
   - **a** a right-angled triangle
   - **b** a rectangle
   - **c** a parallelogram
   - **d** an irregular pentagon

4 Teacher to check. Look for more information than "It is an octagon". Responses could include the word "irregular" and identification of the types of angles.

Mastery

1 2 × large right-angled and isosceles triangles, 1 × medium right-angled and isosceles triangle, 2 × small right-angled and isosceles triangles, 1 square and 1 parallelogram.

2 & 3

Teachers may wish to print off a centimetre-square grid on card as this will make designing the patterns easier. Students could be asked to research the history of the tangram puzzle. Hundreds of designs can be found with a simple online search.

4 Teacher to check. Look for students who are able to accurately represent and identify a range of shapes and who use the correct number of angles in their picture.

UNIT 6: Topic 2 3D shapes

Practice

1 **a** A: triangular prism
   B: square pyramid
   C: sphere
   D: cylinder
   E: cube (accept square prism)
   F: rectangular prism
   G: octagonal prism
   H: cone
   I: triangular pyramid

   b Teacher to check. Answers will vary, e.g. "Objects C, D and H are not polyhedra, because the faces are not all flat."

2 Teacher to check explanations. Likely responses include: "Drawing A is a hexagonal prism. It has two bases."/"Drawing B is a hexagonal pyramid. It has one base."

Challenge

1

	How many faces?	How many edges?	How many vertices?	How many bases?	Shape of base	Shape of side face
a	4	6	4	1	triangle	triangle
b	7	15	10	2	pentagon	rectangle
c	6	12	8	2	square	rectangle
d	10	24	16	2	octagon	rectangle
e	6	10	6	1	pentagon	triangle
f	1	0	0	0	no base	no side faces

2 **a** square pyramid
   b Teacher to check. Answers will vary, e.g. "Because the side faces would be too low to meet together at the top of the pyramid."

Mastery

1 Practical activity.

2 Teachers may wish to make magazines available. Alternatively, students could search online.

UNIT 7: Topic 1 Angles

Practice

1 Teacher to check. Answers will vary, e.g. "Because it is less than a right angle."

2 **a** acute angle: 50°
   b obtuse angle: 100°
   c obtuse angle: 105°
   d acute angle: 75°

3 **a** acute angle: 30°
   b acute angle: 85°
   c right angle: 90°
   d obtuse angle: 100°
   e acute angle: 15°
   f acute angle: 45°

Challenge

1 a 50° **b** 90° **c** 110°
   d 45° **e** 95° **f** 125°

2 Teachers may wish to ask students to explain the strategy they used.
   a Reflex angle: 360° – 80° = 280°
   b Reflex angle: 360° – 115° = 245°

3 Teacher to check and to decide on level of tolerance. (Suggest 2°–3°)

Mastery

1 Angles of the four triangles are:
   a All angles are 60°, total = 180°
   b Angle A = 60°, B = 90°, C = 30°, total = 180°
   c Angle A = 90°, B = 45°, C = 45°, total = 180°
   d Angle A = 90°, B = 55°, C = 35°, total = 180°

2 and 3

Practical activities. Students may find with irregular shapes that the sum of the angles appears not to be 180° (triangle) or 360° (quadrilateral) due to the imprecise nature of the drawing and measuring instruments normally available in a primary school. This could prove to be a good group discussion point.

UNIT 8: Topic 1 Transformations

Practice

1 **a** reflection
   b rotation
   c translation

2 **a** The shape has been translated horizontally.
   b The shape has been reflected horizontally.
   c The shape has been reflected vertically.
   d The shape has been translated diagonally.
   e The shape has been reflected horizontally.
   f The shape has been reflected diagonally.

3 The shape has been reflected horizontally.

Challenge

1

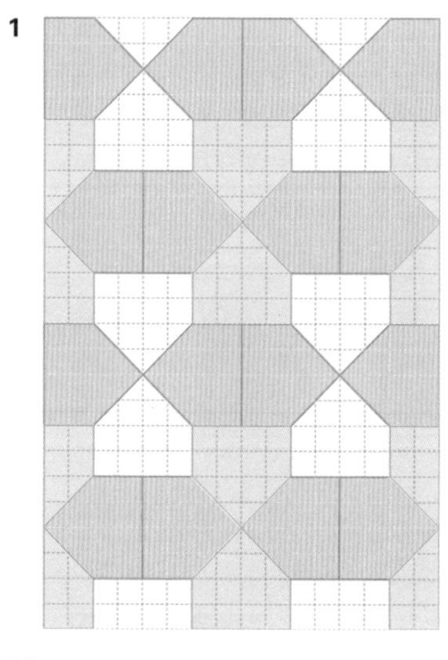

Mastery

1 Practical activity. Teachers may wish to make spare centimetre-square grid paper available for students to practise on.

UNIT 8: Topic 2 Symmetry

Practice

1

a

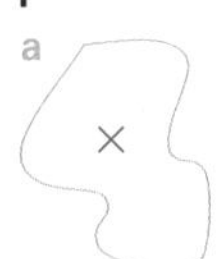

b

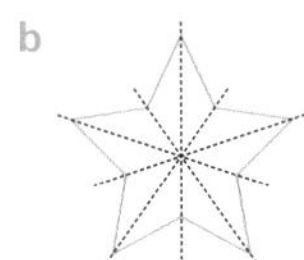

c

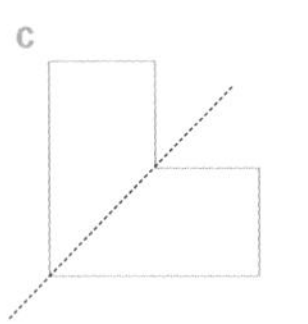

d

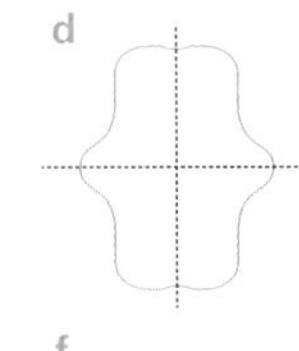

e

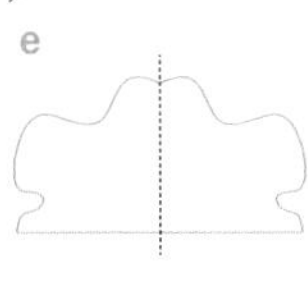

f

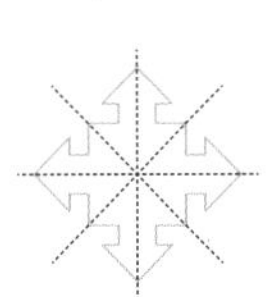

g h

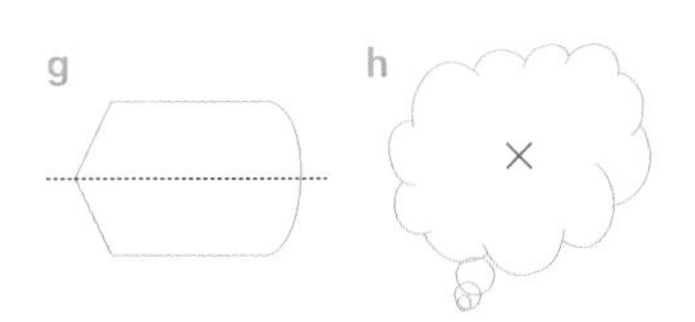

i

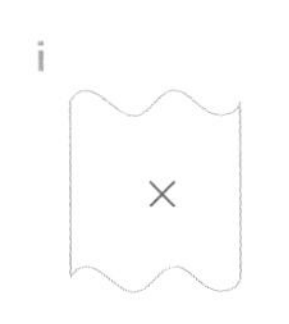

j

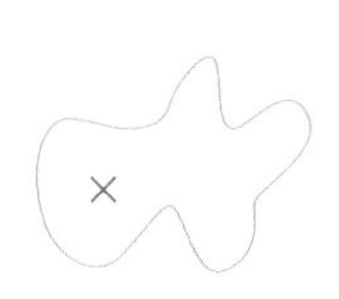

k

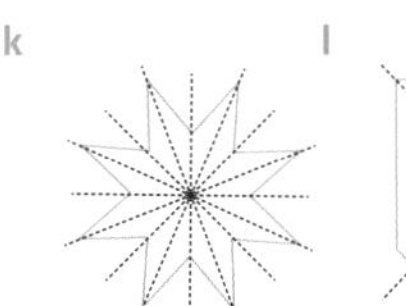

l

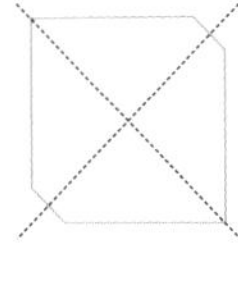

2 **a** 2 **b** 3 **c** 1
d 6 **e** 2 **f** 7

Challenge

1 **a** The letter S has no line of symmetry (but it does have rotational symmetry of order 2).
b Answers will vary. This could become a group or class challenge.

2 Students will find the task easier if they use capital letters. Responses could include: BOOK, CHECK, CHICK, CODEBOOK, COOKBOOK, DECIDED, DIOXIDE, DOB, EXCEEDED, HOODOO

3 Answers could include (depending on how they are written): WOW, OTTO, TOT, TUT, TAT, MOM, YAY.

4 Practical activity.

Mastery

1 The diagonal red stripes do not "meet", which means that the flag is asymmetrical.

2 Practical activity.

UNIT 8: Topic 3 Enlargements

Practice

1

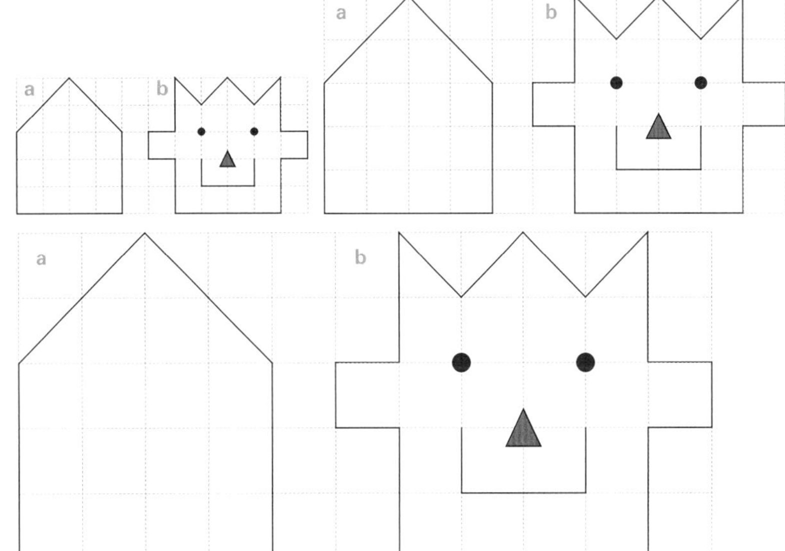

2

ANSWERS

Challenge

1 **a** Enlarge triangle by a scale factor of three.

b Reduce by a scale factor of two.

c Enlarge by a scale factor of two.

d Reduce by a scale factor of three.

Mastery

1 Practical activity.

UNIT 8: Topic 4 Grid references

Practice

1 **a** B3 **b** C3

c (2,3)

2 Teacher to check explanation. The correct answer is Grid A because the grid references refer to the entire area inside a square, whereas on the other two grids, they refer to specific points on the grid.

3 **a** D5 **b** D5

c (4,0)

4 **a** D2 **b** D2

c (0,1)

5 Fourth cross is at (5,1).

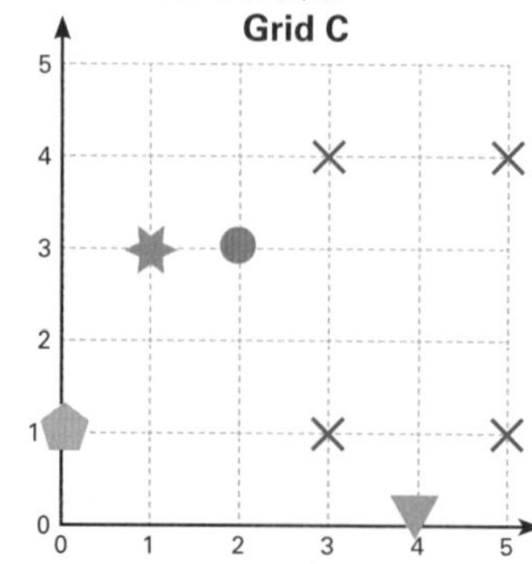

6 **a** (1,5) ↔ (2,3) ↔ (0,1) ↔ (1,5)

b (3,4) ↔ (5,3) ↔ (4,0) ↔ (2,0) ↔ (2,2) ↔ (3,4)

Challenge

1 **a** Answers may vary, as students may draw the letters slightly differently. Possible answer:

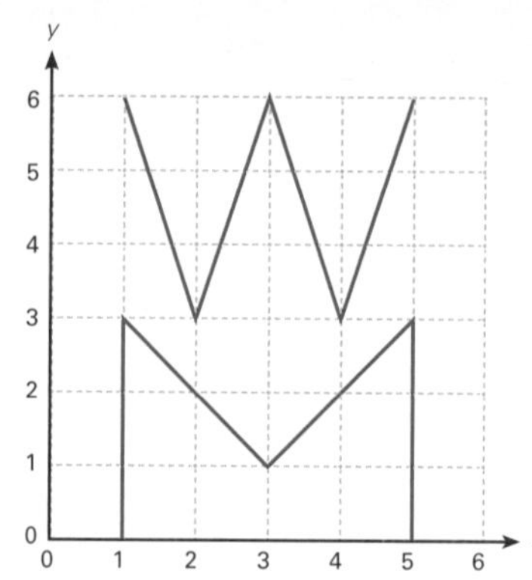

b
- Letter W: (1,6) ↔ (2,3) ↔ (3,6) ↔ (4,3) ↔ (5,6)
- Letter M: (1,0) ↔ (1,3) ↔ (3,1) ↔ (5,3) ↔ (5,0)

2 **a**

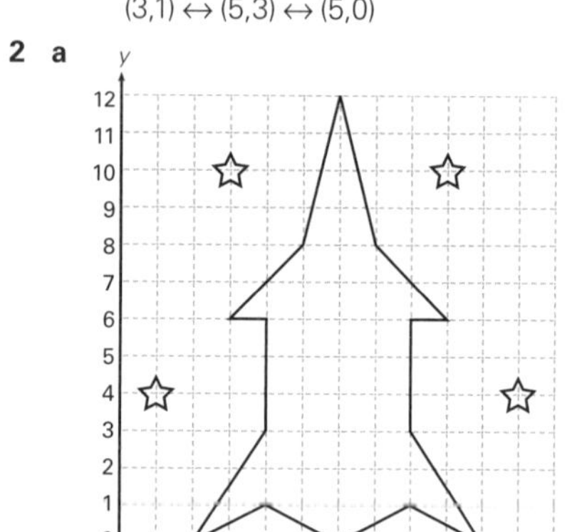

b Students may wish to go the opposite way around the drawing. However, the last coordinate point must be the same as the first.

(6,12) ↔ (5,8) ↔ (3,6) ↔ (4,6) ↔ (4,3) ↔ (2,0) ↔ (4,1) ↔ (6,0) ↔ (8,1) ↔ (10,0) ↔ (8,3) ↔ (8,6) ↔ (9,6) ↔ (7,8) ↔ (6,12)

c Stars are at (1,4), (3,10), (9,10) and (11,4).

Mastery

1 **a** C5 **b** A3, B3, B4 and A4

c Likely response: "Where is the tea cups ride?"

2 Practical activity. If students are motivated to do so, they could draw a larger map on spare grid paper. Students could work on the design in small groups.

UNIT 8: Topic 5 Giving directions

Practice

1 **a**, **b** and **c**

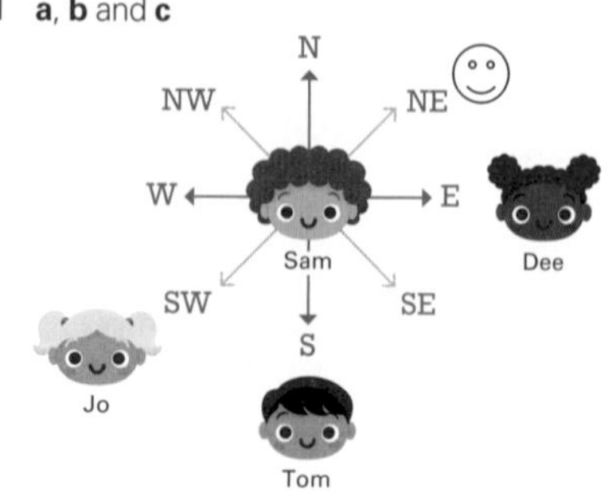

d Jo

2 a west
  b south-east
  c the swim centre
  d Student draws C in A1 below Glenbrook Way and close to Red Road.

Challenge

1 a C1 b E2 c E3

2 a Yass Avenue and Station Street
  b Station Street, Link Way and Showground Way

3 a Green Way, Station Street, Showground Way and Sporting Way
  b Woy Way

4 a south-east b east

5 Teacher to check, e.g.
  a "Go west on Sporting Way and continue west on Showground Way to the shopping centre."
  b "Go north on South Street. Turn west on Showground Way and continue to the shopping centre."

Mastery

1 This could be carried out as a cooperative group activity. Responses could be as detailed or as simple as the teacher/student desires.

2 Students may need some guidance in order to keep the map simple. They could use a scale if desired.

UNIT 9: Topic 1 Collecting and representing data

Practice

1 Students could share their ideas with each other. Example responses:
  a How many books do you read in a week?
  b Do you prefer fiction or non-fiction books?

2 Answers will vary.

3 Graphs should be a reasonable representation of the data that has been collected. Students could read one another's graphs for this purpose.

Challenge

1 a 24
  b Mars
  c Jupiter and Saturn

2 a

Planet	Time to turn once	Rounded time to turn once
Earth	23 hours 56 minutes	24 hours
Mars	24 hours 40 minutes	25 hours
Jupiter	9 hours 54 minutes	10 hours
Saturn	10 hours 40 minutes	11 hours
Uranus	17 hours 14 minutes	17 hours
Nep-tune	16 hours 11 minutes	16 hours

  b Teacher to decide on an acceptable level of accuracy for each section of the graph.

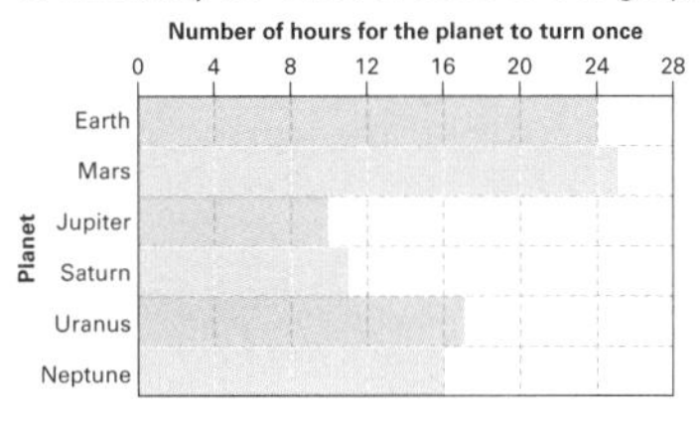

Mastery

1 Students could carry out this task as a cooperative group project.

UNIT 9: Topic 2 Representing and interpreting data

Practice

1

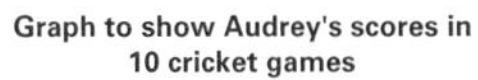

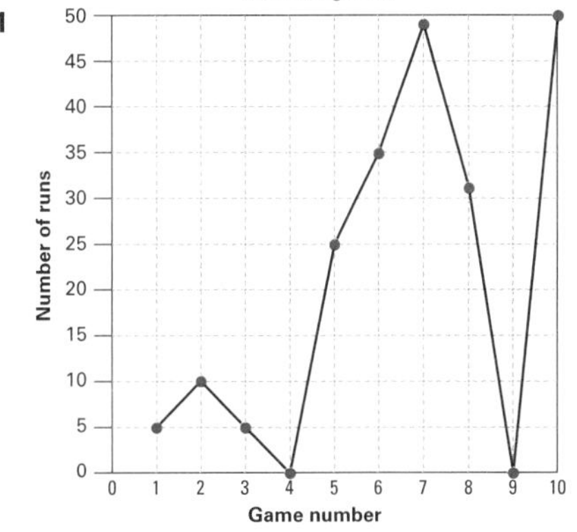

2 a Games 5 and 6
  b Teacher to check. Possible answer: "Her scores dropped from 49 to zero in the two games."
  c Games 7 and 10
  d 210
  e Games 1, 2, 3, 4 and 9. Average score was 21.

3 Answers may vary. Students could share their responses with each other. Possible response:
"As the car's speed increases the thinking distance increases, but the braking distance increases by an even greater amount."

Mastery

1 Practical activity.

Challenge

1 Note, the distances for 110 km/h in the table below are estimates. Students' estimates will vary.

Speed	25 km per hour	50 km per hour	75 km per hour	90 km per hour	110 km per hour
Thinking distance	5 m	15 m	30 m	45 m	65 m
Braking distance	5 m	20 m	45 m	65 m	100 m
Total stopping distance	10 m	35 m	75 m	110 m	165 m

2 Note: Distances for 110 km/h are estimated.

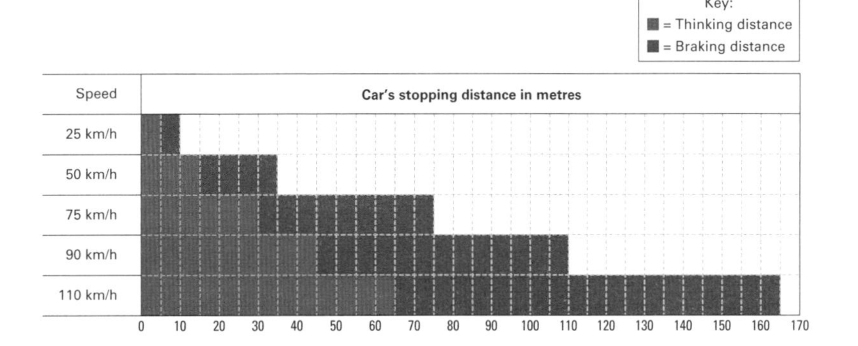

ANSWERS

UNIT 10: Topic 1 Chance

Practice

1 Answers will vary. Students could be asked to justify different values. Teacher to check that the answers given for questions H, I and K are appropriate.

	Description	Value: Decimal	Value: Fraction	Value: %
A	It is *almost certain* that someone will laugh today.	0.9	$\frac{9}{10}$	90%
B	It is *certain* that someone will talk today.	1	1	100%
C	It is *very unlikely* that I will see a movie star this weekend.	0.02	$\frac{2}{100}$	2%
D	It is *unlikely* that I will like every lesson in high school.	0.3	$\frac{3}{10}$	30%
E	It is *very likely* that I will make a spelling mistake this year.	0.8	$\frac{8}{10}$	80%
F	There is *a better than even chance* that someone will sing today.	0.6	$\frac{6}{10}$	60%
G	They say there is *less than an even chance* that it will rain.	0.4	$\frac{4}{10}$	40%
H	It is *impossible* that (Individual responses will vary.)	0	0	0
I	It is *almost impossible* that (Individual responses will vary.)	0.1	$\frac{1}{10}$	10%
J	It is *likely* that I will smile today.	0.7	$\frac{7}{10}$	70%
K	There is an *even chance* that (Individual responses will vary.)	0.5	$\frac{1}{2}$ or $\frac{5}{10}$	50%

2 25%

3 $\frac{1}{4}$

4 0.3

5 $\frac{3}{4}$, 0.75, 75%

Challenge

1 Look for students who recognise that twentieths can easily be converted into equivalent fractions as tenths in order to solve the problems.

a $\frac{2}{20}(\frac{1}{10})$ **b** 0.1 **c** 10%

2 **a** The sections of the spinner should be coloured as follows:
green: 2 red: 2 yellow: 6
blue: 5 purple: 4

b There should be one section left white. This can be expressed as:
fraction: $\frac{1}{20}$ decimal: 0.05
percentage: 5%

3 Look for students who recognise that 8 is one-sixth of the total number of beads in a container. Based on the laws of probability, the total for each colour will be six times the number that Sam took out.

	After 8 have been taken out		My prediction after all 48 have been taken out	
Cup A	blue: 2	white: 6	blue: 12	white: 36
Cup B	blue: 4	white: 4	blue: 24	white: 24
Cup C	blue: 5	white: 3	blue: 30	white: 18
Cup D	blue: 7	white: 1	blue: 42	white: 6

4 **a** There is a one-in-six chance, so the fraction is $\frac{1}{6}$.

b $\frac{1}{6}$ (Every time the dice rolls, there is a $\frac{1}{6}$ chance of it landing on any of the six numbers.)

Mastery

1 Look for students who recognise that the person carried out a total of 100 spins. If the number for each colour is rounded to the nearest 10, the result is red: 10, blue: 50, yellow: 30 and green: 10. This means that the spinner would need to be drawn with $\frac{1}{10}$ each for red and green, half for blue and $\frac{3}{10}$ for yellow.

Students may not choose to be so exact with their designs. The important thing is for them to recognise that red and green will cover significantly less of the design than yellow, and that blue will occupy more than yellow (approximately half of the design).

2 Practical activity. This could be carried out as a cooperative group task.

UNIT 10: Topic 2 Chance experiments

Practice

1 **a** $\frac{1}{2}$ **b** $\frac{5}{6}$

2 **a** 75% **b** 25% **c** 10

3 Practical activity. Students could share their reasons with each other. Possible response: "It all depended on chance."

4 A group discussion would probably correctly conclude that the chance for the "lucky" number had increased from 25% to 40%.

Challenge

1 **a** 0.5, $\frac{1}{2}$ or 50% **b** 0.25, $\frac{1}{4}$ or 25%

2 **a** 10 **b** 10 **c** 5
d 5 **e** 5 **f** 5

3 Answers will vary.

4 Teacher to check. Students could share their reasons with each other. It is unlikely that the predictions were realised but also unlikely that, for example, 20 red cards were chosen. Possible response: "The type of card I picked up depended on chance."

5 A group discussion would probably correctly conclude that it is very unlikely that the same results would happen a second time.

Mastery

1 Answers may vary, e.g. throwing dice or spinning a spinner. Anything in which the chances remain constant.

2 **a** $\frac{9}{10}$ **b** $\frac{1}{10}$

3 Because there are only nine marbles left, there is a slightly better chance ($\frac{1}{9}$) of choosing the white marble. However, there is still $\frac{8}{9}$ chance of not choosing it.

4 The chance has improved from $\frac{1}{9}$ to $\frac{1}{8}$.

5 It is only 100% certain when the white marble is the only one that remains. The chances get better each time but are still no better than 50% when there are two marbles left.

6 **a** $\frac{1}{13}$ **b** $\frac{12}{13}$

7 The chances are far less ($\frac{1}{52}$) and it would probably take longer for the favourite card to appear.